The Green Economy and the Water-Energy-Food Nexus

Robert C. Brears

The Green Economy and the Water-Energy-Food Nexus

Robert C. Brears
Mitidaption
Christchurch, Canterbury, New Zealand

ISBN 978-1-349-84460-9   ISBN 978-1-137-58365-9   (eBook)
DOI 10.1057/978-1-137-58365-9

Library of Congress Control Number: 2017949912

This Palgrave Macmillan imprint is published by Springer Nature
The registered company is Macmillan Publishers Ltd.
The registered company address is: The Campus, 4 Crinan Street, London, N1 9XW, United Kingdom

Acknowledgements

I wish to first thank Rachael Ballard who is not only a wonderful commissioning editor but a visionary that enables books like mine to come to fruition. Second, I would like to thank all the people who have taken time out for an interview, to provide primary materials, and to offer invaluable insights. Without your help this book would not have been possible. Specifically, I wish to thank Minister Melanie Schultz van Haegen (Minister of Infrastructure and the Environment, Netherlands), Kate Auty (Commissioner for Sustainability and the Environment, ACT), Ben van de Wetering (Former Secretary General of the International Commission for the Protection of the Rhine), Marieke Doolaard (CBL Netherlands), Stephen Estes-Smargiassi (Massachusetts Water Resources Authority), Luis S. Generoso (City of San Diego Public Utilities), Amy Kean (New South Wales Department of Industry), Vlada Kenniff (NYC Environmental Protection), Lone Ryg Olsen (Danish Food Cluster), Henning H. Sittel (Effizienz-Agentur NRW), and Mette Hoberg Tønnesen (Ministry of Foreign Affairs Denmark). Finally, I wish to thank mum who has a great interest in the environment and natural resource–related issues and has supported me in this journey of writing the book.

Contents

List of Tables

Introduction

The traditional economic model of employing various types of capital, including human, technological, and natural, to produce goods and services has brought about many benefits including higher living standards and improved human well-being. At the same time, economic growth has resulted in environmental degradation. In addition, the global economic model is confronted by a wide array of trends including rapid population growth, urbanisation, increasing poverty, and inequality as well as climate change resulting in resource scarcity and social challenges. In response, many multi-lateral organisations have called for the development of a green economy that improves human well-being and social equity and reduces environmental degradation.

With rising demand for water, energy, and food, managing the water-energy-food nexus is a key aspect of developing the green economy as the nexus approach recognises the need to use resources more efficiently while seeking policy coherence across the nexus sectors in support of green growth. Nonetheless, the governance of water-energy-food sectors has generally remained separate with limited attention placed on the interactions that exist between them. The result has been narrowly focused actions that have failed to reduce nexus-wide pressures.

The transition towards the green economy will require policy instruments that reduce water-energy-food nexus pressures. In particular, policies will be required to promote social and technological innovations that increase resource efficiency and conservation in order to reduce nexus pressures. Policies that reduce nexus pressures in the green economy can be implemented by fiscal and non-fiscal tools, where fiscal tools include market-based instruments and pricing to encourage resource conservation as well as financial incentives to encourage the uptake of new technologies that reduce nexus pressures. Non-fiscal tools include education and awareness on the need to increase water and energy efficiency across the whole economy as well as reduce food wastage.

However, our understanding of the role governance plays in transitioning towards a green economy that reduces nexus pressures lags significantly behind engineering and physical science knowledge on the interlinking of water, energy, and food. As such, little has been written on the actual implementation of innovative policies at the urban, state, national, and regional levels that promote social and technological change to reduce water-energy-food nexus pressures. In addition, because the transition towards a green economy that reduces nexus pressures requires holistic planning, little has been written on how innovative policies have been developed to integrate the management of water, energy, and food across different sectors of the economy to reduce nexus pressures.

This book, containing case studies, will provide new research on policy innovations at various levels of governance that promote the transition towards a green economy that reduces water-energy-food nexus pressures. In particular, the book will contain case studies that illustrate how cities, states, nations, and regions of differing climates, lifestyles, and income levels have implemented fiscal and non-fiscal policy innovations to reduce water-energy-food nexus pressures and achieve green growth.

Specifically, the book contains nine case studies of leading cities, states, nations, and regions that have implemented a variety of fiscal and non-fiscal policy tools to create interdependencies and synergies between the nexus systems while reducing trade-offs between the systems in the development of a green economy. Data for the case studies are collected from interviews, primary materials, and insights from government departments or agencies in charge of various aspects of the water-energy-food nexus. The nine case studies are New York City and Singapore; Massachusetts and Ontario; Denmark and Korea; and the Colorado River Basin, Murray-Darling River Basin, and the Rhine River Basin.

The case studies are chosen for the following reasons. New York City is considered a leader in policy innovations that reduce resource consumption and enhance social equity; Singapore, faced with land constraints, is implementing a variety of policies to be more resource efficient and less constrained by its lack of natural resources; Massachusetts is considered one of the most innovative states in the United States, known for high-tech research on resource efficiency and renewable energy; Ontario is Canada's largest province by population size and a key economic region of North America, yet at the same time considered progressive in its environmental and resource efficiency policies; Denmark is considered by many international rankings as a leader in terms of resource efficiency and conservation; Korea, faced with limited natural resources, has become a leading proponent of green growth in Asia and globally; the Colorado River Basin is faced with enormous climatic pressures while being the main source of water, energy, and food for tens of millions of people in the United States; the Murray-Darling River Basin is the food bowl of Australia, yet faces nexus pressures with the basin being home to the majority of Australians; and the Rhine River Basin, which flows through the heart of Europe, is the epicentre of Europe's largest economies and therefore under multiple nexus pressures.

The book's chapter synopsis is as follows:

Chapter 1 will introduce readers to the numerous challenges faced by the current economic model followed by an overview of the green economy and how the green economy can facilitate sustainable development.

Chapter 2 discusses the components of the water, energy, and food nexus before discussing the various interactions between the nexus's sectors. Following which the chapter outlines the various impacts of nexus pressures before discussing the benefits of reducing these pressures.

Chapter 3 provides readers with a review of the various fiscal and non-fiscal tools available to reduce nexus pressures and achieve a green economy along with mini-case examples of policies being implemented around the world.

Chapter 4 provides readers with a case study on the green economy and the water-energy food nexus in New York City.

Chapter 5 provides readers with a case study on the green economy and the water-energy food nexus in Singapore.

Chapter 6 provides readers with a case study on the green economy and the water-energy food nexus in Massachusetts.

Chapter 7 provides readers with a case study on the green economy and the water-energy food nexus in Ontario.

Chapter 8 provides readers with a case study on the green economy and the water-energy food nexus in Denmark.

Chapter 9 provides readers with a case study on the green economy and the water-energy food nexus in Korea.

Chapter 10 provides readers with a case study on the green economy and the water-energy food nexus in the Colorado River Basin.

Chapter 11 provides readers with a case study on the green economy and the water-energy food nexus in the Murray-Darling River Basin.

Chapter 12 provides readers with a case study on the green economy and the water-energy food nexus in the Rhine River Basin.

Chapter 13 provides readers with a series of best practices that can be implemented by other locations around the world attempting to reduce water-energy-food nexus pressures and achieve a green economy.

1

The Green Economy

Introduction

Economic growth typically refers to an increase in the level of goods and services produced by an economy, as estimated by measures including gross domestic product (GDP). It involves the combination of different types of capital to produce goods and services including produced capital, such as machinery; human capital, such as knowledge and skills; natural capital, such as raw materials; and social capital, including institutions. While economic growth has produced many benefits including raising living standards and improving quality of life around the world, it has also resulted in the depletion of natural resources and degradation of ecosystems. As such, there has been much debate on whether or not it is possible to achieve economic growth without unsustainably degrading the environment.[1]

1.1   Challenges to the Global Economy

The key challenges to the global economy include a variety of issues including population and economic growth, rapid urbanisation, environmental degradation, and climate change.[2]

© The Author(s) 2018
R.C. Brears, *The Green Economy and the Water-Energy-Food Nexus*,
DOI 10.1057/978-1-137-58365-9_1

Population Growth

The world's population is projected to increase by more than one billion people within the next 15 years, reaching 8.5 billion in 2030 and to increase up to 9.7 billion in 2050 and 11.2 billion by 2100. More than half of that population growth is expected to occur in Africa, while Asia will add 0.9 billion people between 2015 and 2050. At the country level, half of the world's population growth is expected to occur in just nine countries: India, Nigeria, Pakistan, Democratic Republic of Congo, Ethiopia, Tanzania, the United States, Indonesia, and Uganda.[3]

Rapid Urbanisation

In 2014, 54 percent of the world's population resided in urban areas, compared to 30 percent in 1950. This is projected to rise to 66 percent by 2050. Currently, the most urbanised regions of the world are North America (82 percent), Latin America and Caribbean (80 percent), and Europe (73 percent). In comparison, Africa and Asia are mainly rural with 40 and 48 percent of their populations urban respectively; however, these two regions are urbanising faster than any other region: by 2050, 56 percent of Africa's population will be urban while in Asia 64 percent will be living in urban areas. In terms of population size, increasing population growth and urbanisation are combined expected to add 2.5 billion people to the world's urban population by 2050, with nearly 90 percent of this increase expected to be in Africa and Asia. The number of mega-cities—cities with more than 10 million inhabitants—is also projected to increase: in 1990, there were ten of these cities. This has risen to 28 today, and by 2030, the number of mega-cities will reach 41.[4]

Rapid Economic Growth

Over the period of 2014–2050, it is projected that the world's economy will grow at an average of over 3 percent per annum, doubling in size by 2037 and nearly tripling by 2050. At the same time, global economic power will shift away from the established advanced economies in North

America, Western Europe, and Japan, with India projected to become the second-largest economy by 2050 in purchasing power parity (PPP). Meanwhile, emerging economies such as Mexico and Indonesia are likely to be larger than the United Kingdom and France by 2030 (in PPP terms). In terms of the rate of growth, Nigeria and Vietnam could be the fastest-growing large economies between now and 2050 at 5 percent per annum, in comparison to China's 3–4 percent growth rate and advanced economies' rate of 1.5–2.5 percent.[5] This will increase demand for water, energy, food, and natural resources, resulting in global material resource consumption projected to double by 2050.[6]

Air Pollution and Climate Change

The World Health Organization (WHO) estimates that 92 percent of the world's population lives in places where air quality exceeds WHO limits. Around 3 million deaths a year are linked to exposure to outdoor air pollution. Indoor air pollution can be just as deadly with 6.5 million deaths in 2012 associated with indoor and outdoor air pollution together. Nearly 90 percent of air pollution–related deaths occur in low- and middle-income countries, of which 94 percent are due to non-communicable diseases, particularly cardiovascular diseases, stroke, chronic obstructive pulmonary disease, and lung cancer.[7] In Japan, prenatal exposure to outdoor air pollution has been associated with behavioural problems relating to attention and delinquent or aggressive behaviour at age eight in a nationally representative sample across the country.[8]

Deforestation

Forests provide food, wood energy, shelter, fodder, and fibre while harbouring biodiversity. However, the world's forests continue to diminish as populations increase and forest land is converted to agriculture and other uses. In fact, 80 percent of all deforestation is a result of commercial agriculture particularly in developing countries.[9] Since 1990, nearly 130 million hectares of forest—around the same size as South Africa—have been lost, with Africa and South America having the highest net annual loss of forests over the period 2010–2015, with 2.8 and 2 million hectares respectively.[10,11,12]

Ecosystem Degradation

The Millennium Ecosystem Assessment found that around 60 percent of the ecosystems evaluated in the assessment, including 70 percent of regulating and cultural services, were found to be degraded or used unsustainably. Ecosystem services that have been degraded over the past 50 years include capture fisheries, water supply, waste treatment and detoxification, water purification, natural hazard protection, regulation of air quality, regulation of regional and local climate, regulation of erosion, spiritual fulfilment, and aesthetic enjoyment. The use of two ecosystem services—capture fisheries and freshwater—is beyond levels that can sustain even current demands: around a quarter of all commercial fish stocks are over-harvested, and from 5 to even 25 percent of global freshwater use exceeds long-term accessible supplies and is now met through either engineered water transfers or over-abstraction of groundwater supplies.[13] In New Zealand, Lake Rotorua's ecosystem services value is calculated to be NZD 94–138 million per annum with potential damage costs of eutrophication calculated at $14–49 million.[14]

Soil Degradation

Between 10 and 20 percent of land globally is already degraded,[15] leading to soil degradation, which is the decline in soil quality caused by improper use, usually for agricultural, pastural, industrial, or urban purposes. It encompasses physical, chemical, and biological deterioration. Examples of soil degradation include loss of organic matter; decline in soil fertility; erosion; adverse changes in salinity; and the effects of toxic chemicals, pollutants, or excessive flooding.[16] In Europe, the estimated number of contaminated sites adds up to 1.5 million, of which more than 300,000 have been identified.[17] Meanwhile, in sub-Saharan Africa, a combination of poor land management, a growing population, and a changing climate is leading to soil degradation, with economic losses estimated to be $68 billion per year.[18]

Poverty and Inequality

Despite global extreme poverty levels declining from 12.4 percent in 2012 to 10.17 percent in 2013, the goal of ending extreme poverty worldwide by 2030 is at risk. Even the high economic growth rates of the past two decades will be insufficient to end extreme poverty by this date. For instance, the gains against poverty have not been uniform across regions: in 2013, sub-Saharan Africa accounted for more of the poor—389 million—than all other regions combined. This is in contrast to 1990 when half of the poor were living in East Asia and the Pacific. In addition, within-country inequality is greater now than 25 years ago; for example, in South Africa the top income share roughly doubled over 20 years to levels comparable to those observed in the United States.[19]

Trade and Globalisation

The expansion of international trade and foreign investment has led to the lifecycles of products becoming 'displaced'. Today, the extraction, production, consumption, and recycling and/or final disposal of a product often take place at great distances from one another and/or in completely different countries. This has allowed companies to take advantage of cost savings by outsourcing to countries where the necessary factors of production are plentiful and cheaper (labour, energy, land, etc.). This process has displaced the potential environmental impacts associated with each stage of a product's lifecycle, shifting the environmental burden between regions of the world.[20]

Changing Consumption Patterns

One of the main demand drivers for natural resources is the expansion of the global middle class, which in turn leads to growth in consumer expenditure. Over the next two decades more than three billion will reach middle-class status, the majority of which will come from emerging markets.[21] In 53 emerging and developing economies, there was an increase of 37 percent in the number of households earning more than USD

10,000 between 2010 and 2015. This has led to middle-class households buying more consumer goods and increasing demand for raw materials across a range of natural resources including energy, metals, and water. For example, in China, the percentage of households owning a refrigerator has increased from nearly 40 percent in 2000 to around 75 percent in 2015. Similarly, the number of households owning a washing machine increased from 40 percent in 2000 to 65 percent in 2015.[22]

Technological Development and Innovation

Technology and innovation play an important role in improving resource productivity and mitigating the negative environmental impacts of resource use and material consumption. However, their role is complex with technology and innovation having the power to alter both supply and demand of primary and secondary raw materials. For instance, efficient production processes can reduce waste and alleviate pressure on resources. At the same time, new technologies have the potential to generate additional environmental pressures.[23]

Climate Change

Climate change affects all regions of the world with polar ice shields melting causing sea levels to rise, while in some regions extreme weather events and rainfall will become more common while others will experience more extreme heat waves and droughts. For instance, climate change–induced higher temperatures will increase demand for electricity for cooling purposes, while hotter summers will increase risks of heat stress as well as an increase in disease. More frequent intense rainfall will increase the likelihood of flooding. Water scarcity will increase too with higher temperatures and lower rainfall reducing soil moisture and groundwater supplies. Drier conditions in some areas will also result in more frequent droughts. Lower water flows will raise water temperatures, reducing water quality (e.g. through increased algae growth). Rising sea levels will increase the risk of erosion, inundation, and saltwater intrusion. Warmer temperatures will impact biodiversity with some species' habitats altered, resulting in increased risk of localised extinction. Warmer temperatures will provide favourable conditions for exotic species as well as increase the risks of diseases and pests spreading, affecting both fauna and flora.[24]

1.2 The Green Economy

The United Nations Environment Programme (UNEP) defines the green economy as one that results in 'improved human well-being and social equity, while significantly reducing environmental risks and ecological scarcities'. In its simplest form, a green economy is low carbon, resource efficient, and socially inclusive. In this type of economy, growth in income and employment are driven by both public and private investments that reduce carbon emissions, enhance resource efficiency, and prevent the loss of biodiversity and ecosystem services. A key component of this economy is that economic development views natural capital as a key economic asset and as a source of public benefit. The overall aim of a transition towards a green economy is to enable economic growth and investment while increasing environmental quality and social inclusiveness.[25] The green economy approach is a shift away from the short-term understanding of environmental considerations as a cost factor that constrains economic growth and reduces competitiveness. Instead, it views these considerations as fundamental to the long-term sustainability of economic growth.[26] Overall, the green economy is one that results in 'improved human well-being and social equity, while significantly reducing environmental risks and ecological scarcities'.[27]

Green Growth

The benefits of economic growth were gained at the expense of the environment. The green paradigm indicates that the previous approach to growth, such as polluting and degrading the environment first and then cleaning up and restoring the environment afterwards, must be suspended. Instead, a new path should advocate sustainable development that protects the environment. Green growth has emerged as a new development paradigm to respond to the traditional unsustainable energy and carbon-intensive models that are based on economic growth without consideration for the environment.[28] According to the Organization for Economic Co-operation and Development (OECD), green growth is about fostering economic growth and development while ensuring that natural assets continue to provide the resources and environmental services on which our well-being relies. 'To do this, green growth must catalyse investment and innovation which will underpin sustained growth and give rise to new economic opportunities'.[29] Green growth is relevant to developed countries which need to retrofit their resource-consuming industries and lifestyles and to developing countries that can avoid copying damaging development pathways. They can 'leapfrog' old solutions and adopt new technologies and ideas. Meanwhile, in developed countries, the challenge of transitioning towards the green economy will be to change lifestyles and reduce consumption of natural resources to sustainable levels. In developing countries, the challenge will be to stimulate economic growth so the green economy coincides with sustainable development.[30] A variety of characteristics attributed to green growth include:

- More effective use of natural resources in economic growth
- Valuing ecosystems
- Inter-generational economic policies
- Increased use of renewable sources of energy
- Protection of vital assets from climate-related disasters
- Reduced waste of resources

Objectives of the Green Economy

Overall, despite the definitions of green economy and green growth differing across different international organisations (summarised in Table 1.1) the terms are broadly characterised by three objectives:

- Improving resource-use efficiency: a green economy is one that is efficient in its use of energy, water, and other material inputs.
- Ensuring ecosystem resilience: it also protects the natural environment, its ecosystems, and ecosystem flows.
- Enhancing social equity: it promotes human well-being and a fair burden sharing across societies.[31]

Table 1.1 Definitions of green economy and green growth

International agency	Definition
European Environment Agency	A green economy is one in which socio-economic systems are organised in ways that enable society to live well within planetary boundaries. Resource efficiency is increased in the green economy, which is complemented with enhancing ecosystem resilience and people's well-being.
OECD	Green growth means fostering economic growth and development, while ensuring that natural assets continue to provide the resources and environmental services on which our well-being relies. To do this, it must catalyse investment and innovation which will underpin sustained growth and give rise to new economic opportunities.
UNDP	New growth poles that can potentially contribute to economic recovery, decent job creation, and reduce threats of food, energy, ecosystem, and climate crises, which have disproportionate impacts on the poor.
UNEP	A green economy is one that results in improved human well-being and social equity, while significantly reducing environmental risks and ecological scarcities. A green economy can be thought of as one which is low carbon, resource efficient and socially inclusive.
World Bank	Green growth—that is, growth that is efficient in its use of natural resources, clean in that it minimises pollution and environmental impacts, and resilient in that it accounts for natural hazards and the role of environmental management and natural capital in preventing physical disasters. And this growth needs to be inclusive.

EEA. 2014. Resource-efficient green economy and EU policies. Available: http://www.eea.europa.eu/publications/resourceefficient-green-economy-and-eu; OECD. 2017. *What is green growth and how can it help deliver sustainable development?* [Online]. Available: http://www.oecd.org/general/whatisgreengrowthandhowcanithelpdeliversustainabledevelopment.htm; UNDP. 2012. *Green economy: A transformation to address multiple crises* [Online]. Available: http://www.undp.org/content/undp/en/home/presscenter/articles/2009/06/25/green-economy-a-transformation-to-address-multiple-crises.html; UNEP. 2012. *About: Green growth* [Online]. Available: http://www.unep.org/rio20/About/GreenEconomy/tabid/101541/Default.aspx; World Bank. 2012. Inclusive green growth: The pathway to sustainable development. Available: http://elibrary.worldbank.org/doi/book/10.1596/978-0-8213-9551-6

Improving Resource-Use Efficiency

With a rising global population increasing demand for water, energy, food, and natural resources, the result will be an increase in global material resource consumption levels.[32] Therefore, there is the need to sustain economic growth in the longer term while keeping negative environmental impacts under control and preserving natural resources.[33] By improving resource efficiency it means economies can use the Earth's limited resources in a sustainable way to create more with less while reducing residual waste to close to zero.[34] A key aspect of improving resource-use efficiency is decoupling economic activity from consumption and environmental impacts. There are two dimensions to decoupling: resource decoupling means reducing the rate of use of resources per unit of economic activity and impact decoupling means maintaining economic output while reducing the negative environmental impact of any economic activities undertaken. Combined, relative decoupling of resources or impacts means the growth rate of the resources used or environmental impacts is lower than the economic growth rate, so that resource productivity is rising. Absolute reductions of resource use happen when the growth rate of resource productivity exceeds the growth rate of the economy.[35]

Ensuring Ecosystem Resilience

An ecosystem is a dynamic complex of plant, animal, and microorganism communities and the non-living environment interacting as a functional unit. An ecosystem can be the distribution of organisms, soil types, drainage basins, or even depth in a body of water. Ecosystem services are the benefits humans obtain from ecosystems. These include provisioning services such as water and food; regulating services such as flood and disease control; cultural services such as spiritual, recreational, and cultural benefits; and supporting services such as nutrient cycling that maintain the conditions for life on Earth.[36] Ecosystem resilience meanwhile refers to the capacity of an ecosystem to recover from disturbance or withstand ongoing pressures. It is the measure of how well an ecosystem can tolerate disturbance without collapsing into a different state. An ecosystem's ability to absorb or recover from impacts and its rate of recovery depend on the biology and ecology of its component species or habitats; the conditions of these individual components; the nature, severity, and duration of the impacts; and the degree to which potential impacts have been removed or reduced. If any limitations exist, the capacity of the ecosystem to absorb impacts without changing will be lower than optimal and recovery will take much longer or even fail.[37]

Enhancing Social Equity

The integration of the social aspect, or human well-being, in the green economy is fundamental given the importance of basic resources—food, water, energy, and materials—as well as ecosystem services for human subsistence needs.[38] Regarding green jobs and inclusiveness, job opportunities in an inclusive green economy will be significant but they need to be identified early and the education and skills training needed to fill them will need to be a priority for any government engaging in a transition towards this new economy. Overall, the extent of the creation of green jobs will be seen as an indicator of competitiveness: the higher an economy's creation of green jobs, the better it is posed to compete in a future where economies are designed to produce wealth and income without creating environmental risks and resource scarcities.[39] Meanwhile, on the issue of equity and natural resources, the green economy depends on the capacity of disadvantaged groups to organise collectively; engage in contestation, advocacy, and bargaining; and be part of broader coalitions of change. To facilitate active citizenship, the green economy's governance arrangements need to be sensitive to the issues of diversity, representation, and space for negotiation and ensure policies are not dominated by narrow or elite interests. Governments can also cultivate an enabling environment for participation and empowerment through education and training as well as institutionalising accountability mechanisms and basic rights and freedoms of association, expression, information, and redress.[40]

The Green Economy and Sustainable Development

The concept of the green economy does not replace that of sustainable development; instead, it can be understood as a way to achieve sustainable development,[41] which is defined by the Brundtland Commission's *Our Common Future* as development that meets the needs of the present without compromising the ability of future generations to meet their own needs. Consisting of three pillars, sustainable development seeks to achieve, in a balanced manner, economic development, social development, and environmental protection.[42]

Economic Sustainability

Weak sustainability assumes that natural capital and manufactured capital are essentially substitutable and that there are no essential differences between the kinds of well-being they create: the only thing that matters is the total value of the aggregate stock of capital, which should be at least maintained or ideally increased for the sake of future generations. In addition, in the weak form of sustainability the economy will continually generate technical solutions to environmental problems caused by increased production levels of goods and services. In contrast, the strong sustainability view sees natural capital as a complex system that consists of evolving biotic and abiotic elements that interact in ways that determine the ecosystem's capacity to provide ecosystem services. In this context, strong sustainability proponents state that natural capital is non-substitutable as natural capital cannot be reproduced and its destruction is irreversible, in contrast to manufactured capital which can be reproduced and restored. In addition, due to the lack of knowledge about the functioning of ecosystems, society cannot be certain about the effects on human well-being from destroying natural capital. As such, acknowledging irreversibility and uncertainties should lead to the implementation of the precautionary principle regarding the use of natural capital in economic growth.[43]

Social Sustainability

An unjust society is unlikely to be sustainable in environmental or economic terms. Rather, social tensions are likely to undermine the recognition
by citizens of both their environmental rights and duties relating to environmental degradation.[44] Therefore, a better understanding of sustainable
development's concept of social sustainability is critical for reconciling the
competing demands of the society-environmental-economic tripartite.[45]
There are five interconnected equity principles of social sustainability:

- *Intergeneration equity*: Which is equity between generations where
  future generation's standards of living should not be disadvantaged by
  the activities of the current generation's standard of living.
- *Intragenerational equity*: Which is equity amongst the current generation and can be achieved through widespread political participation by
  citizens.
- *Geographical equity* (trans-frontier responsibility): Whereby local policies should be geared towards resolving local and global environmental
  problems as political/administrative boundaries are frequently used to
  shield polluters from prosecution in other jurisdictions.
- *Procedural equity*: Regulatory systems should be devised to ensure
  transparency as it is critical that people have the right to access environmental information on activities that have both local and global
  impacts.
- *Interspecies equity*: This notion places the survival of other species on an
  equal basis to the survival of humans. This is to reflect the critical
  importance of preserving ecosystems and maintaining biodiversity for
  human survival. Specifically, humans have an obligation to ensure ecosystems are not degraded beyond their regenerative capacity.[46,47,48,49,50]

Environmental Sustainability

Environmental sustainability is the ability to maintain and enhance the
qualities that are valued in the physical environment, for instance,

maintaining the living conditions for people and other species. Specifically, environmental sustainability aims to protect the integrity of natural eco-systems and the various ecosystem services necessary for human survival including clean air and purified water. There are four types of ecosystem services that nature provides: Provisioning services are products obtained from ecosystems, for example, food and water; regulating services are benefits obtained from the regulation of ecosystem processes, for instance, climate regulation (maintenance of temperatures), water regulation (flood protection), and water purification; supporting services which are necessary for the continuation of the three above-mentioned ecosystem services types, for example, soil formation, nutrient cycling, and primary production; and cultural services, which are non-material benefits obtained from ecosystems such as religious, spiritual, recreational, and educational.[51]

Nonetheless, while governments agreed to frame the green economy as an important tool for sustainable development, uncertainty and ambiguity remain around how governments should now apply this concept.[52]

Notes

1. Tim Everett, M. I., Gian Paolo Ansaloni and Alex Rubin. 2010. Economic growth and the environment. *Defra Evidence and Analysis Series* [Online]. Available: https://www.gov.uk/government/uploads/system/uploads/attachment_data/file/69195/pb13390-economic-growth-100305.pdf.
2. Rasul, G. 2016. Managing the food, water, and energy nexus for achieving the Sustainable Development Goals in South Asia. *Environmental Development*, 18, 14–25.
3. UN Department of Economic and Social Affairs, P. D. 2015. World population prospects: The 2015 revision, key findings and advance tables. Available: https://esa.un.org/unpd/wpp/publications/files/key_findings_wpp_2015.pdf.
4. UN Department of Economic and Social Affairs, P. D. 2014. World urbanisation prospects: The 2014 revision, highlights. Available: https://esa.un.org/unpd/wup/publications/files/wup2014-highlights.pdf.

5. PWC. 2015. The World in 2050: Will the shift in global economic power continue? Available: http://www.pwc.com/gx/en/issues/the-economy/assets/world-in-2050-february-2015.pdf.

6. OECD. 2016. Policy guidance on resource efficiency. Available: http://www.oecd.org/environment/waste/Resource-Efficiency-G7-2016-Policy-Highlights-web.pdf.

7. WHO. 2016. *WHO releases country estimates on air pollution exposure and health impact* [Online]. Available: http://www.who.int/mediacentre/news/releases/2016/air-pollution-estimates/en/.

8. Yorifuji, T., Kashima, S., Diez, M. H., Kado, Y., Sanada, S. & Doi, H. 2017. Prenatal exposure to outdoor air pollution and child behavioral problems at school age in Japan. *Environment International,* 99, 192–198.

9. Leblois, A., Damette, O. & Wolfersberger, J. 2017. What has Driven Deforestation in Developing Countries Since the 2000s? Evidence from New Remote-Sensing Data. *World Development,* 92, 82–102.

10. FAO. 2016a. State of the world's forests 2016: Forests and agriculture: Land use challenges and opportunities. Available: http://www.fao.org/3/a-i5588e.pdf.

11. FAO. 2016b. *World deforestation slows down as more forests are better managed* [Online]. Available: http://www.fao.org/news/story/en/item/326911/icode/.

12. Kissinger, G., M. Herold, V. De SY., 2012. Drivers of Deforestation and Forest Degradation: A Synthesis Report for REDD+ Policymakers. Available: https://www.gov.uk/government/uploads/system/uploads/attachment_data/file/65505/6316-drivers-deforestation-report.pdf.

13. Millennium Ecosystem Assessment. 2005. *Ecosystems and human well-being: synthesis,* Washington, DC, Island Press.

14. Mueller, H., Hamilton, D. P. & Doole, G. J. 2016. Evaluating services and damage costs of degradation of a major lake ecosystem. *Ecosystem Services,* 22, Part B, 370–380.

15. ELD Initiative. 2015. Report for policy and decision makers: Reaping economic and environmental benefits from sustainable land management. Available: http://www.eld-initiative.org/fileadmin/pdf/ELD-pm-report_05_web_300dpi.pdf.

16. New South Wales Government Office of Environmental Heritage. 2017. *Soil degradation* [Online]. Available: http://www.environment.nsw.gov.au/soildegradation/.

17. EEA. 1999. Environment in the European Union at the turn of the century Available: http://www.eea.europa.eu/publications/92-9157-202-0.

18. The Montpellier Panel. 2014. No ordinary matter: Conserving, restoring and enhancing Africa's soils. Available: http://ag4impact.org/wp-content/uploads/2014/12/MP_0106_Soil_Report_LR1.pdf.

19. World Bank. 2016. Poverty and shared prosperity 2016: Taking on inequality. Available: https://openknowledge.worldbank.org/bitstream/handle/10986/25078/9781464809583.pdf.

20. OECD. 2015. Material resources, productivity and the environment. *OECD Green Growth Studies* [Online]. Available: http://www.oecd.org/env/waste/material-resources-productivity-and-the-environment-9789264190504-en.htm.

21. Ernst and Young. 2013. Hitting the sweet spot: The growth of the middle class in emerging markets. Available: http://www.ey.com/Publication/vwLUAssets/Hitting_the_sweet_spot/$FILE/Hitting_the_sweet_spot.pdf.

22. Euromonitor International. 2016. *What drives demand for natural resources* [Online]. Available: http://blog.euromonitor.com/2016/12/what-drives-demand-natural-resources.html.

23. OECD. 2015. Material resources, productivity and the environment. *OECD Green Growth Studies* [Online]. Available: http://www.oecd.org/env/waste/material-resources-productivity-and-the-environment-9789264190504-en.htm.

24. New Zealand Ministry for the Environment. 2017. *How climate change could affect New Zealand* [Online]. Available: http://www.mfe.govt.nz/climate-change/how-climate-change-affects-nz/climate-change-impacts.

25. UNEP. 2011b. Towards a Green Economy: Pathways to Sustainable Development and Poverty Eradication. Available: http://web.unep.org/greeneconomy/resources/green-economy-report.

26. EEA. 2014. Resource-efficient green economy and EU policies Available: http://www.eea.europa.eu/publications/resourceefficient-green-economy-and-eu.

27. UNEP. 2011b. Towards a Green Economy: Pathways to Sustainable Development and Poverty Eradication. Available: http://web.unep.org/greeneconomy/resources/green-economy-report.

28. K-Water Institute and World Water Council. 2015. Water and green growth: Beyond the theory for sustainable future. Available: http://www.worldwatercouncil.org/fileadmin/world_water_council/documents/mailing/mail_wwf7_documents/Water_and_Green_Growth_vol_1.pdf.

29. OECD 2011. Towards green growth.

30. GWP. 2012. Water in the green economy. Available: http://www.gwp.org/Global/ToolBox/Publications/Perspective%20Papers/03%20Water%20in%20the%20Green%20Economy%20(2012).pdf.
31. EEA. 2014. Resource-efficient green economy and EU policies. Available: http://www.eea.europa.eu/publications/resourceefficient-green-economy-and-eu.
32. OECD. 2016. Policy guidance on resource efficiency. Available: http://www.oecd.org/environment/waste/Resource-Efficiency-G7-2016-Policy-Highlights-web.pdf.
33. OECD. 2015. Material resources, productivity and the environment. *OECD Green Growth Studies* [Online]. Available: http://www.oecd.org/env/waste/material-resources-productivity-and-the-environment-9789264190504-en.htm.
34. European Commission. 2011. Roadmap to a resource-efficient Europe. Available: http://eur-lex.europa.eu/legal-content/EN/TXT/PDF/?uri=CELEX:52011DC0571&from=EN.
35. UNEP. 2011a. Decoupling natural resource use and environmental impacts from economic growth. Available: http://www.unep.org/resourcepanel/Portals/50244/publications/Decoupling_Factsheet_English.pdf.
36. Alcamo, J., Bennett, E. M. & Assessment, M. E. 2003. *Ecosystems and Human Well-being: A Framework for Assessment*, Island Press.
37. Australian Government Great Barrier Reef Marine Park Authority. 2009. Great Barrier Reef outlook report 2009. Available: http://www.gbrmpa.gov.au/__data/assets/pdf_file/0018/3843/OutlookReport_Full.pdf.
38. EEA, 2014. Resource-efficient green economy and EU policies. https://www.eea.europa.eu/publications/resourceefficient-green-economy-and-eu.
39. UNEP. 2015. Uncovering pathways towards an inclusive green economy. Available: http://web.unep.org/greeneconomy/sites/unep.org.greeneconomy/files/publications/ige_narrative_summary_web.pdf.
40. UNRISD. 2012. Social dimensions of green economy (Research and policy brief). Available: http://www.unrisd.org/80256B3C005BCCF9/search/84781D22707BF025C1257A0E0054A320?OpenDocument.
41. EEA. 2014. Resource-efficient green economy and EU policies Available: http://www.eea.europa.eu/publications/resourceefficient-green-economy-and-eu.
42. Brundtland Commission 1987. Our common future: Report of the World Commission on Environment and Development. *UN Documents Gathering a Body of Global Agreements*.

43. Pelenc, J., Ballet, J. & Dedeurwaerdere, T. 2015. Weak sustainability versus strong sustainability. *Brief for GSDR United Nations* [Online]. Available: https://sustainabledevelopment.un.org/content/documents/6569122-Pelenc-Weak%20Sustainability%20versus%20Strong%20Sustainability.pdf.
44. Haughton, G. 1999. Environmental justice and the sustainable city. *Journal of Planning Education and Research,* 18, 233–243.
45. Vallance, S., Perkins, H. C. & Dixon, J. E. 2011. What is social sustainability? A clarification of concepts. *Geoforum,* 42, 342–348.
46. Haughton, G. 1999. Environmental justice and the sustainable city. *Journal of Planning Education and Research,* 18, 233–243.
47. Curwell, S. & Cooper, I. 1998. The implications of urban sustainability. *Building Research & Information,* 26, 17–28.
48. Lieberherr-Gardiol, F. 2008. Urban sustainability and governance: issues for the twenty-first century. *International Social Science Journal,* 59, 331–342.
49. Cuthill, M. 2010. Strengthening the 'social' in sustainable development: Developing a conceptual framework for social sustainability in a rapid urban growth region in Australia. *Sustainable Development,* 18, 362–373.
50. Lieberherr-Gardiol, F. 2008. Urban sustainability and governance: issues for the twenty-first century. *International Social Science Journal,* 59, 331–342.
51. The Economics of Ecosystems and Biodiversity. 2015. *Ecosystem services* [Online]. Available: http://www.teebweb.org/resources/ecosystem-services/.
52. UNDESA. 2012. A guidebook to the green economy. Available: https://sustainabledevelopment.un.org/content/documents/GE%20Guidebook.pdf.

References

Alcamo, J., E.M. Bennett, and M.E. Assessment. 2003. *Ecosystems and human well-being: A framework for assessment.* Washington, DC: Island Press.
Australian Government Great Barrier Reef Marine Park Authority. 2009. Great Barrier Reef outlook report 2009. http://www.gbrmpa.gov.au/__data/assets/pdf_file/0018/3843/OutlookReport_Full.pdf
Brundtland Commission. 1987. Our common future: Report of the World Commission on Environment and Development. *UN Documents Gathering a Body of Global Agreements.*

Curwell, S., and I. Cooper. 1998. The implications of urban sustainability. *Building Research & Information* 26: 17–28.

Cuthill, M. 2010. Strengthening the 'social' in sustainable development: Developing a conceptual framework for social sustainability in a rapid urban growth region in Australia. *Sustainable Development* 18: 362–373.

EEA. 1999. Environment in the European Union at the turn of the century. http://www.eea.europa.eu/publications/92-9157-202-0

———. 2014. Resource-efficient green economy and EU policies. http://www.eea.europa.eu/publications/resourceefficient-green-economy-and-eu

ELD Initiative. 2015. Report for policy and decision makers: Reaping economic and environmental benefits from sustainable land management. http://www.eld-initiative.org/fileadmin/pdf/ELD-pm-report_05_web_300dpi.pdf

Ernst and Young. 2013. Hitting the sweet spot: The growth of the middle class in emerging markets. http://www.ey.com/Publication/vwLUAssets/Hitting_the_sweet_spot/$FILE/Hitting_the_sweet_spot.pdf

Euromonitor International. 2016. *What drives demand for natural resources* [Online]. http://blog.euromonitor.com/2016/12/what-drives-demand-natural-resources.html

European Commission. 2011. Roadmap to a resource efficient Europe. http://eur-lex.europa.eu/legal-content/EN/TXT/PDF/?uri=CELEX:52011DC0571&from=EN

FAO. 2016a. State of the world's forests 2016: Forests and agriculture: Land use challenges and opportunities. http://www.fao.org/3/a-i5588e.pdf

———. 2016b. *World deforestation slows down as more forests are better managed* [Online]. http://www.fao.org/news/story/en/item/326911/icode/

GWP. 2012. Water in the green economy. http://www.gwp.org/Global/ToolBox/Publications/Perspective%20Papers/03%20Water%20in%20the%20Green%20Economy%20(2012).pdf

Haughton, G. 1999. Environmental justice and the sustainable city. *Journal of Planning Education and Research* 18: 233–243.

K-Water Institute and World Water Council. 2015. Water and green growth: Beyond the theory for sustainable future. http://www.worldwatercouncil.org/fileadmin/world_water_council/documents/mailing/mail_wwf7_documents/Water_and_Green_Growth_vol_1.pdf

Kissinger, G., M. Herold, and V. De Sy. 2012. Drivers of deforestation and forest degradation: A synthesis report for REDD+ policymakers. https://www.gov.uk/government/uploads/system/uploads/attachment_data/file/65505/6316-drivers-deforestation-report.pdf

Leblois, A., O. Damette, and J. Wolfersberger. 2017. What has driven deforestation in developing countries since the 2000s? Evidence from new remote-sensing data. *World Development* 92: 82–102.

Lieberherr-Gardiol, F. 2008. Urban sustainability and governance: Issues for the twenty-first century. *International Social Science Journal* 59: 331–342.

Millennium Ecosystem Assessment. 2005. *Ecosystems and human well-being: Synthesis*. Washington, DC: Island Press.

Mueller, H., D.P. Hamilton, and G.J. Doole. 2016. Evaluating services and damage costs of degradation of a major lake ecosystem. *Ecosystem Services* 22 (Part B): 370–380.

New South Wales Government Office of Environmental Heritage. 2017. *Soil degradation* [Online]. http://www.environment.nsw.gov.au/soildegradation/

New Zealand Ministry for the Environment. 2017. *How climate change could affect New Zealand* [Online]. http://www.mfe.govt.nz/climate-change/how-climate-change-affects-nz/climate-change-impacts

OECD. 2011. *Towards green growth*. Paris: OECD.

———. 2015. Material resources, productivity and the environment. *OECD Green Growth Studies* [Online]. http://www.oecd.org/env/waste/material-resources-productivity-and-the-environment-9789264190504-en.htm

———. 2016. Policy guidance on resource efficiency. http://www.oecd.org/environment/waste/Resource-Efficiency-G7-2016-Policy-Highlights-web.pdf

———. 2017. *What is green growth and how can it help deliver sustainable development?* [Online]. http://www.oecd.org/general/whatisgreengrowthandhowcanithelpdeliversustainabledevelopment.htm

Pelenc, J., J. Ballet, and T. Dedeurwaerdere. 2015. Weak sustainability versus strong sustainability. *Brief for GSDR United Nations* [Online]. https://sustainabledevelopment.un.org/content/documents/6569122-Pelenc-Weak%20Sustainability%20versus%20Strong%20Sustainability.pdf

PWC. 2015. The world in 2050: Will the shift in global economic power continue? http://www.pwc.com/gx/en/issues/the-economy/assets/world-in-2050-february-2015.pdf

Rasul, G. 2016. Managing the food, water, and energy nexus for achieving the sustainable development goals in South Asia. *Environmental Development* 18: 14–25.

The Economics of Ecosystems and Biodiversity. 2015. *Ecosystem services* [Online]. http://www.teebweb.org/resources/ecosystem-services/

The Montpellier Panel. 2014. No ordinary matter: Conserving, restoring and enhancing Africa's soils. http://ag4impact.org/wp-content/uploads/2014/12/MP_0106_Soil_Report_LR1.pdf

Tim Everett, M.I., Gian Paolo Ansaloni, and Alex Rubin. 2010. Economic growth and the environment. *Defra Evidence and Analysis Series* [Online]. https://www.gov.uk/government/uploads/system/uploads/attachment_data/file/69195/pb13390-economic-growth-100305.pdf

UN Department of Economic and Social Affairs, Population Division. 2014. World urbanization prospects: The 2014 revision, highlights. https://esa.un.org/unpd/wup/publications/files/wup2014-highlights.pdf

———. 2015. World population prospects: The 2015 revision, key findings and advance tables. https://esa.un.org/unpd/wpp/publications/files/key_findings_wpp_2015.pdf

UNDESA. 2012. A guidebook to the green economy. https://sustainabledevelopment.un.org/content/documents/GE%20Guidebook.pdf

UNDP. 2012. *Green economy: A transformation to address multiple crises* [Online]. http://www.undp.org/content/undp/en/home/presscenter/articles/2009/06/25/green-economy-a-transformation-to-address-multiple-crises.html

UNEP. 2011a. Decoupling natural resource use and environmental impacts from economic growth. http://www.unep.org/resourcepanel/Portals/50244/publications/Decoupling_Factsheet_English.pdf

———. 2011b. Towards a green economy: Pathways to sustainable development and poverty eradication. http://web.unep.org/greeneconomy/resources/green-economy-report

———. 2012. *About: Green growth* [Online]. http://www.unep.org/rio20/About/GreenEconomy/tabid/101541/Default.aspx

———. 2015. Uncovering pathways towards an inclusive green economy. http://web.unep.org/greeneconomy/sites/unep.org.greeneconomy/files/publications/ige_narrative_summary_web.pdf

UNRISD. 2012. *Social dimensions of green economy.* Research and Policy Brief. http://www.unrisd.org/80256B3C005BCCF9/search/84781D22707BF025C1257A0E0054A320?OpenDocument

Vallance, S., H.C. Perkins, and J.E. Dixon. 2011. What is social sustainability? A clarification of concepts. *Geoforum* 42: 342–348.

WHO. 2016. *WHO releases country estimates on air pollution exposure and health impact* [Online]. http://www.who.int/mediacentre/news/releases/2016/air-pollution-estimates/en/

World Bank. 2012. Inclusive green growth: The pathway to sustainable development. http://elibrary.worldbank.org/doi/book/10.1596/978-0-8213-9551-6
———. 2016. Poverty and shared prosperity 2016: Taking on inequality. https://openknowledge.worldbank.org/bitstream/handle/10986/25078/9781464809583.pdf
Yorifuji, T., S. Kashima, M.H. Diez, Y. Kado, S. Sanada, and H. Doi. 2017. Prenatal exposure to outdoor air pollution and child behavioral problems at school age in Japan. *Environment International* 99: 192–198.

2

The Green Economy and the Water-Energy-Food Nexus

Introduction

Water, energy, and food are inextricably linked where the actions in one sector influence the actions in other sectors. For instance, food production requires water; water extraction, treatment, and distribution require energy; and energy production requires water. As such, researchers and policymakers have increasingly emphasised the importance of managing the 'water-energy-food nexus' (WEF nexus)[1] as the nexus approach can support the transition towards the green economy, which aims to use resources efficiently while seeking greater policy coherence across the nexus sectors.[2] However, the governance of WEF nexus sectors remains separate with limited attention given to the interactions that exist between them.[3] The result has been narrowly focused actions, investments, and policies that fail to reduce nexus pressures.[4]

© The Author(s) 2018
R.C. Brears, *The Green Economy and the Water-Energy-Food Nexus,*
DOI 10.1057/978-1-137-58365-9_2

2.1   The Components of the WEF Nexus

The individual components of the WEF nexus sectors are as follows:

Water

Water, unlike any other natural resource, affects every aspect of society and the environment and is essential for human well-being. Specifically, water is embedded in all aspects of development including food security, health, and poverty reduction and in sustaining economic growth in agriculture, industry, and energy generation.[5] As such, the transition towards the green economy requires not only the conservation of water resources but also the finding of new and innovative economic growth and social development opportunities that embrace the sustainable management of water resources. A key component of creating the green economy is ensuring water security for all users and uses, both human and natural,[6] where water security is *the capacity of a population to safeguard sustainable access to adequate quantities of and acceptable quality water for sustaining*

Table 2.1 Synergies between water and green growth

Characteristics of green growth	Characteristics of water security
More effective use of natural resources in economic growth	Ensure enough water for social and economic development
Valuing ecosystems	Ensure adequate water for maintaining ecosystems
Inter-generational economic policies	Sustainable water availability for future generations
Protection of vital assets from climate-related disasters	Balance the intrinsic value of water with its uses for human survival and well-being
Reduce waste of resources	Harness the productive power of water, maintain water quality, and avoid pollution and degradation

UN-Water. 2013. Water security and the global water agenda; Global Water Partnership. 2012. Water in the green economy. Available: http://www.gwp.org/Global/ToolBox/Publications/Perspective%20Papers/03%20Water%20in%20the%20Green%20Economy%20(2012).pdf *livelihoods, human well-being, and socio-economic development, for ensuring protection against water-borne pollution and water-related disasters, and for preserving ecosystems in a climate of peace and political stability.* The synergies between green growth and water security are summarised in Table 2.1. Overall, ensuring water security in the green economy can be achieved by:

- Creating policy instruments that promote complementary benefits (economic, environmental, social)
- Developing fiscal instruments that give a price to environmental goods
- Strengthening institutional arrangements that enable the management of water across sectoral silos and even political/administrative boundaries
- Developing financial instruments that share risks between governments and investors and make new water technology affordable
- Developing skills that support the sustainable management of water in the green economy
- Establishing information and monitoring systems that set targets, define trajectories, and monitor progress on water efficiencies
- Developing innovative plans that increase water productivity, protect groundwater and surface water resources, and ensure adequate levels of water quality

Energy

Energy underpins economic growth in all countries irrespective of levels of development with it being key in the production of all types of goods and services including food and the provision of water. Meanwhile, energy fuel enables the movement of manufactured goods, people, services, and ideas. As such, energy use and economic growth are closely linked with a clear linear relationship between energy consumption and wealth, measured in GDP of nations.[7] Improving the environmental performance of energy is considered by the OECD as a cornerstone of any attempt towards green growth. In a green economy, the energy sector's role is replacing fossil fuel with low-carbon options in addition to

promoting energy efficiency and renewable energy with the overall aim
of decoupling economic growth from energy demand.[8] However, shortfalls in investment in clean energy can have severe consequences for
energy security as well as long-term economic growth and climate
change mitigation,[9] where energy security is defined by the International
Energy Agency (IEA) as the 'uninterrupted availability of energy sources
at an affordable price'. It is important to note that there are two aspects
of energy security: in the long term, energy security deals with timely
investments to supply energy in line with economic development and
sustainable environmental needs, while in the short-term energy security

Table 2.2 The energy sector in the green economy

Element	Description
Recognising economic costs of environmental damage and poorly managed natural resources	A failure to address environmental concerns and not managing natural resources pose risks to long-term economic growth, through: • rising prices caused by resource scarcity • increased burden of environmental damage caused by the conventional use of fossil fuels • negative consequences of climate change on human well-being and impairment of human health caused by pollution
Innovation to achieve environmental and economic objectives	• Innovations help decouple environmental damage from economic growth while increasing productivity and jobs • Innovations can create forms of energy that impose fewer environmental costs and improve efficiency in use as prices rise
Synergies between environmental and productivity growth objectives	Improved resource productivity and energy efficiency through innovation or use of energy technology or processes supports the decoupling of economic growth from environmental damage and resource degradation
Opportunities for new markets and industries	• The transition towards green growth in the energy sector will require new technologies, fuel sources, processes, and services • Increased demand from consumers and businesses for environmentally friendly products, services, and production processes in the energy sector

OECD. 2011b. OECD green growth studies: Energy. Available: https://www.oecd.org/greengrowth/greening-energy/49157219.pdf

focuses on the ability of the energy sector to react promptly to sudden changes in the supply-demand balance.[10] Overall there are four key elements that provide the economic rationale for placing the energy sector on a green growth path that also achieves energy security, which are summarised in Table 2.2.

Food

The food sector has been successful in providing for increasing global demand for a long period of time. Agricultural productivity has exceeded that in many other sectors and has exceeded the population growth rate.[11] In addition to providing food, the agricultural sector has the potential to alleviate poverty in developing countries as on average the contribution of agriculture to raising the incomes of the poorest is estimated to be at least 2.5 times higher than that of non-agriculture sectors.[12] In the transition towards the green economy, the food and agricultural sector will reduce negative environmental effects while increasing productivity and farmer incomes all the while ensuring food security for all, where food security is defined by the Food and Agriculture Organization (FAO) as 'all people at all times have physical, social and economic access to sufficient, safe and nutritious food to meet their dietary needs and food preferences for an active, healthy life'.[13] A key aspect of achieving food security is recognising that while intensifying crop production can boost the food security of millions of people around the world, increasing food production can contribute to problems including land degradation, water pollution, and depletion of water resources, all of which in turn threaten food security. Ensuring food security in the green economy can be achieved by farming practices and technologies that simultaneously:

- maintain and increase farm productivity and profitability, while ensuring the provision of food and ecosystem services on a sustainable basis;
- reduce negative externalities till positive ones are achieved;
- rebuild ecological resources (i.e. soil, water, air, and biodiversity natural capital assets) by reducing pollution and using resources more efficiently.[14]

2.2   Interactions Across the WEF Nexus

The following section identifies and quantifies the interactions and feedbacks between the components of the WEF nexus.

Water-Energy/Energy-Water

Water is required to produce nearly all forms of energy. For primary fuels water is required in resource extraction, irrigation of biofuel feedstock, fuel refining and processing, and transport. The IEA estimates that global freshwater withdrawals for energy production in 2010 were 583 billion cubic metres, some 15 percent of the world's total water withdrawals.[15] By 2035 it is projected that freshwater withdrawals for energy production will increase to 20 percent.[16] By 2040 the IEA projects global energy demand to increase by 37 percent with the world's energy supply mix divided into almost four equal parts: oil, gas, coal, and low-carbon sources. Regarding fossil-fuel-based energy, increased oil use for transportation and petrochemicals will see global demand for oil increase from 90 million barrels per day in 2013 to 104 million barrels per day in 2040; demand for natural gas will increase by more than half over the same time period—the fastest rate among fossil fuels—while global demand for coal will grow by 15 percent. How much water is required to meet increased demand for fossil-fuel-based energy depends on whether the world follows a business-as-usual approach towards energy efficiency: Following a business-as-usual approach, the IEA projects water demand for energy to be 35 percent higher than 2010, compared to a more energy-efficient future requiring 20 percent more water. Regarding low-carbon sources, global production of liquid biofuels has expanded from 16 billion litres in 2000 to more than 100 billion litres in 2011.[17] However, biofuel has significant impacts on water resources because of its water requirements during crop growth (photosynthesis), and water use in biorefineries.[18] For instance, one study investigated the water requirements of China meeting its biofuel development plans for 2020 in which the country will aim to produce 12 million metric tonnes of biofuels by this date, of which 10 million metric tonnes will be comprised of bioethanol and the remaining two million tonnes being biodiesel. To meet this goal the study found the water requirements ranged between 31.9 km^3 and 71.7 km^3; the latter is the equivalent to the total annual discharge of the Yellow River.[19]

Demand for electricity is projected to increase by 70 percent between now and 2035.[20] Because thermal power generation and hydropower, which accounts for 80 and 15 percent of global electricity generation respectively, is water-intensive, the estimated 70 percent increase in electricity production would translate into a 20 percent increase in freshwater withdrawals[21]: in power generation, water provides cooling for thermal power plants while in hydropower stations its movement is harnessed for electricity production. Water withdrawals per unit of electricity generated are highest for fossil fuel (coal-, gas-, and oil-fired plants operating on a steam cycle) and nuclear power plants with once-through cooling at 75,000–450,000 litres per megawatt-hour (l/MWh), while combined-cycled gas turbines with wet-cooling systems have water withdrawals of 570–1000 l/MWh.[22] Increasingly, energy production affects water quality via contamination by fluids that contain pollutants or physical alteration of the nature environment. In addition, water used for cooling in power plants is often returned to rivers or lakes at a high temperature, which decreases the oxygen content in the water causing thermal pollution and damages ecosystems (Table 2.3).[23,24]

Electricity is needed to power pumps that abstract (from ground and surface sources), transport, distribute, and collect water. The amount of electricity required depends on the distance, or depth, of the water source. Water treatment processes that convert various types of water—fresh, brackish, saline, and waste—into water fit for specific use require electricity and at times heat. In 2014, around 4 percent of global electricity consumption was used in the water sector. By 2040, the amount of energy used in the water sector is projected to double due to increases in desalination and wastewater treatment services, particularly in emerging economies. Already 9 percent of Saudi Arabia's annual electricity consumption is attributed to groundwater pumping and desalination while other countries in the Arabian Gulf consume 5–12 percent or more of

Table 2.3 Uses of water for energy and potential water quality impacts

Primary energy production	Uses	Potential water quality impacts
Oil and gas	• Drilling, well completion, and hydraulic fracturing • Injection into reservoir in secondary and enhanced oil recovery • Oil sands and in situ recovery • Upgrading and refining into products	Contamination by tailings seepage, fracturing fluids in surface and groundwater
Coal	• Cutting and dust suppression in mining and hauling • Washing to improve coal quality • Re-vegetation of surface • Long-distance transport	Contamination by tailings seepage, mine drainage, or produced water impacting surface and groundwater
Biofuels	• Irrigation for feedstock crop growth • Wet milling, washing, and cooling in fuel conversion process	• Contamination of surface and groundwater by runoff containing fertilisers, pesticides and sediments • Wastewater produced by refining
Power generation	Uses	Potential water quality impacts
Thermal (fossil fuel, nuclear energy, and bioenergy)	• Boiler feed (water used to generate steam or hot water) • Cooling for steam-condensing • Pollutant scrubbing using emission-controlling equipment	• Thermal pollution by cooling water discharge (into surface water) • Impact on aquatic ecosystems • Air emissions that pollute surface water downwind • Discharge of boiler (boiler feed that contains suspended solids)

(continued)

Table 2.3 (continued)

Concentrating solar power and geothermal	• System fluids or boiler feed • Cooling for steam-condensing	• Thermal pollution of surface water • Impact on aquatic ecosystems
Hydropower	• Electricity generation • Storage in reservoir (for operating hydroelectric dams)	• Alteration of water temperatures, flow volume/timing, and aquatic ecosystems • Evaporative losses from reservoir

IEA. 2012a. Water for energy. Available: http://www.worldenergyoutlook.org/media/weowebsite/2012/WEO_2012_Water_Excerpt.pdf

total electricity consumption for desalination.[25] Moving forward, it is projected that by 2040, 16 percent of energy consumption in the Middle East region will be related to water supply.[26,27]

The construction of dams and reservoirs for electricity generation can also affect water quality, which in turn impacts aquatic ecosystems through the altering of temperatures that can damage ecosystems. For instance, many fish species as well as frogs, crabs, shrimps, and molluscs cannot survive in waters that are 19 °C or warmer as the change in temperature reduces the amount of dissolved oxygen available for species. Other effects include trout eggs unable to hatch in warm water, and, even if they did, they would not survive for long as aquatic juveniles are the least tolerant to abrupt temperature changes. Thermal pollution can also increase the susceptibility of aquatic organisms to parasites, toxins, and pathogens, making them vulnerable to various diseases.[28] If thermal pollution occurs for a long time it can cause large-scale bacteria and plant growth, leading to algae bloom that will in turn result in even less oxygen in the water. Finally, sediment build-up from dams and reservoirs can increase turbidity of water bodies, which not only can reduce water depths resulting in increased water temperatures, but it can prevent the growth of plants, reducing the ability of fish to find food and detect prey, therefore increasing stress.[29]

Table 2.4 Risks and impacts within the water-energy/energy-water nexus

	Risks	Impacts
Water-related risks to energy security	Shifts in availability and quality due to natural or human reasons	Reduced availability of supply and reliance on more expensive forms of generation
		Possibility of economic pricing of water increasing costs of energy production
		Reduced availability of water for fuel extraction and processing stages leading to reduced output
	Increase in energy demand for water production, treatment, and distribution	Strains on energy system and reduced efficiencies from different demand profiles for water and energy
Energy-related risks to water security	Limited or unreliable access to affordable energy for extracting water	Disruption in water supply to end-users or diversion of resources away from other core activities, for example agriculture
	Re-allocation of water resources from other end-users to energy	Changes in delivery costs of water due to fluctuating costs of energy inputs
	Contamination of water resources due to energy extraction and transformation processes	Water resources, including drinking water, becomes contaminated, requiring additional treatment

IRENA. 2015b. *Renewable energy in the water, energy and food nexus* [Online]. Available: http://www.irena.org/documentdownloads/publications/irena_water_energy_food_nexus_2015.pdf

An overall summary of the risks and impacts of the water-energy/energy-water nexus is summarised in Table 2.4.

Water-Food/Food-Water

Agriculture accounts for 70 percent of all water withdrawn. Annual global agricultural water consumption includes crop water consumption for food, fibre, and seed production plus evaporation losses from the soil and open water associated with agriculture, for example rice fields, irrigation canals, and reservoirs. By 2050, the world will require 60 percent more food produce to maintain current consumption patterns.[30,31] This will result in the volume of global water withdrawn for irrigation increasing from 2.6 billion cubic kilometres in 2005–2007 to 2.9 billion cubic kilometres in 2050.[32] The increase in demand for limited water supplies for agricultural production will likely place pressure on water-intensive food producers to seek alternative supplies, often leading to inter-sectoral competition for limited water resources.[33] China has faced increased demand for food due to a rising population that has increased by around 1 percent over 1978–2014 and rising per capita GDP, which has increased by 9.7 percent over the same period. To meet this demand, China's agricultural water withdrawal and cultivated land has increased 3.2 percent (from 378 cubic kilometres to 390 cubic kilometres) and 5.4 percent respectively between 2000 and 2015. This has led to environmental degradation with 11 percent of the plain regions experiencing groundwater overdraft, resulting in groundwater tables declining, land subsidence, and intrusion of sea water.[34] Meanwhile, in much of Pakistan, the overexploitation of groundwater has led to the country's canal irrigation network becoming severely deteriorated.[35]

The increase in agricultural production will impact the quality of water resources due to non-point source pollution (Table 2.5). Key problems include sediment runoff that causes siltation problems; nutrient runoff with nitrogen and phosphorous being key pollutants found in agricultural runoff having been applied to farmland in several ways including as a fertiliser, animal manure, and municipal wastewater; microbial runoff from livestock or use of excreta as fertiliser; and chemical runoff from pesticides, herbicides, and other agrichemicals contaminating surface and groundwater.[36,37]

Irrigation salinity is the rise in saline groundwater and the build-up of salt in the soil surface in irrigated areas. Inefficient irrigation or applying more water than the plants can use means this excess water leaks past the root zone to groundwater. This can lead to the water table rising, which mobilises salt that has accumulated in the soil layers. When the saline water table rises to within two metres of the surface, evaporation concentrates salt at the surface. As the soil becomes waterlogged vegetation and crops die because they have limited access to oxygen. Soil saturation can also be compounded by periods of heavy rainfall. The effects of salt

Table 2.5 Agricultural impacts on water quality

Agricultural activity	Impacts	
	Surface water	Groundwater
Tillage/ploughing	Sediments and turbidity carry phosphorous and pesticides Siltation of river beds Loss of habitat, spawning grounds	
Fertilising	Runoff of nutrients, including phosphorous, leads to eutrophication	Leaching of nitrate to groundwater
Manure spreading	Can lead to high levels of contamination of receiving waters by microorganisms, phosphorous, and nitrate causing eutrophication	Contamination of groundwater by nitrogen
Pesticides	Runoff of pesticides contaminates surface water, causing dysfunction of ecological system due to growth inhibition and reproductive failure	Some pesticides may leach into groundwater, causing human health problems from contaminated wells
Feedlot	Contamination of surface water with microorganisms and residues of veterinary drugs, contamination by metals in urine and faeces	Potential leaching of nitrogen, metals etc. to groundwater
Irrigation	Runoff of salts leads to salinisation of surface water, runoff of fertilisers and pesticides to surface waters with ecological damage, bioaccumulation in fish species	Contamination of groundwater with salts and nutrients
Clear cutting	Erosion of land, leading to turbidity in rivers, siltation of bottom habitat. Disruption and change of hydrological regime with loss of streams and decreasing flow in dry periods, concentration of nutrients and contaminants	Disruption of hydrological regime often leads to increased surface runoff and decreased groundwater recharge

UN-Water. 2015. Wastewater management: A UN-Water analytical brief. Available: http://www.unwater.org/fileadmin/user_upload/unwater_new/docs/UN-Water_Analytical_Brief_Wastewater_Management.pdf on plants and soil include increased difficulty for plants to extract water; excess sodium can cause leaf burn, dead patches, and even defoliation; and a reduction in uptake of nutrients including iron, while high sodium can exclude potassium.[38,39] The southwest region of Bangladesh is facing salinity intrusion. Over the period 2000–2009, saline water intrusion increased up to 15 kilometres north of the coast and during the dry season reached up to 160 kilometres inland. To reduce salinity in the area, farmers have become dependent more on fertiliser, pesticides, and irrigation equipment to increase their crop yields. In addition, farmers have replaced their traditional cropping practices including mixed cropping and crop rotation with conventional farming practices of mono-cropping. However, this has led to the further deterioration of the soil quality and fertility. In addition, it has increased labour costs as well as fertiliser and pesticide expenses.[40]

Fertiliser runoff can lead to eutrophication of surface waterways. While eutrophication occurs naturally, it is normally associated with excessive inputs of phosphorous (P) and nitrogen (N) where impairment is measured as the area of surface water not suitable for designated uses such as irrigation. Eutrophication has many negative effects on aquatic ecosystems, the most obvious consequence being the increased growth of algae and aquatic weeds that interfere with the use of waters for agriculture as well as recreation, industry, and drinking.[41] In marine ecosystems, algae blooms—red or brown tides—can release toxins and cause anoxia when oxygen is consumed as dead algae decompose. The resulting coastal eutrophication can cause shellfish poisoning in humans and increased mortality rates in marine mammals. If too much oxygen is removed the water body develops a 'dead zone', which is an area that can no longer support fish, shellfish, or most other aquatic life. One of the most well-known dead zones is found in the Gulf of Mexico, which is fed by the nitrate-rich Mississippi River and can fluctuate in size from 3000 to 8000 square miles. Other dead zones include areas in the Baltic Sea, the Adriatic Sea, the Gulf of Thailand, and the Yellow Sea. In freshwater, blooms of cyanobacteria (blue-green algae) is a prominent symptom of eutrophication. These blooms contribute to a wide range of water-related problems including summer fish kills, foul odours, and unpalatable drinking water. Toxins released from these blooms can kill livestock and may pose serious

Table 2.6 Risks and impacts within the water-food/food-water nexus

	Risks	Impacts
Water-related risks to food security	Increased variability in water availability due to climatic and non-climate trends	Changes in supply of food products leading to higher price volatility, compounded by regional concentration of food production activities
	Impact of water quality on food production and consumption	Utilisation of poor-quality water along different stages of the food supply chain can have negative impacts, for example soil degradation, accumulation of contaminants within the food chain
Food-related risks to water security	Impacts of agricultural activities on water resources	Use of external inputs for agriculture and food production can lead to water pollution affecting all users, human and natural
	Poorly regulated agricultural foreign direct investment	Increased agricultural land leasing, when poorly regulated, can lead to expanded use of local water resources with potential local socio-economic impacts
	Water resource over-utilisation due to food security ambitions	Pursuit of food security ambitions can strain water resources, leading to depletion in freshwater reserves

IRENA. 2015a. Renewable energy in the water, energy and food nexus. Available: http://www.irena.org/DocumentDownloads/Publications/IRENA_Water_Energy_Food_Nexus_2015.pdf

health hazards to humans.[42] For instance, nitrates from nitrogen fertilisers and runoff from livestock can infiltrate drinking water with high concentrations leading to 'blue baby disease' in infants. In blue baby disease, nitrate ions weaken the blood's capacity to carry oxygen.[43]

An overall summary of the risks and impacts of the water-food/food-water nexus is summarised in Table 2.6.

Food-Energy/Energy-Food

The food sector currently accounts for around 30 percent of global energy consumption.[44] Agricultural energy consumption at the farm level can be divided into direct and indirect energy use, with direct energy referring to the fuel or electricity used to power farm activities and indirect energy referring to the fertilisers, chemicals, and other agricultural inputs produced off site. Typically, the majority of energy used is in direct operations; for example, in the United States, the agricultural sector consumes around 1.74 percent of total US primary energy consumption with about 60 percent of that consumed directly. The total amount and form of energy consumed vary based on a farm's principal commodity. For example, cotton production requires larger amounts of electricity to operate irrigation pumps while dryland wheat production does not.[45] Energy is required in the operating of groundwater pumps. For instance, in Saudi Arabia, energy consumed in groundwater pumping comprises less than 10 percent of the total energy inputs for food production; however, a transition to desalinated water or brackish water will significantly increase the percentage of energy required due to the additional treatment required.[46] Fertilisers (nitrogen, phosphate, and potash) and chemical pesticides, fungicides, and herbicides—all of which are used to increase food production in all regions of the world—require energy in their production, distribution, and transport process. Fertilisers form the largest of these energy inputs to agriculture, while pesticides are the most energy-intensive agricultural input (on a per-kilogramme basis of chemical).[47] Total fertiliser nutrient consumption is likely to rise by over 200 million tonnes in 2018, 25 percent higher than recorded in 2008 and global use of nitrogen, which is by far the largest fertiliser base, is projected to rise 1.4 percent each year through 2018.[48] However, because the current nitrogen fertiliser production is based on fossil fuel energy, nitrogen fertiliser inputs will contribute significantly to the total greenhouse gas emissions of agricultural output.[49] Energy is also required to produce, transport, and distribute food. Increased transport networks, transnational companies, and supply chains and growing retail markets mean an increasing range of food from around the world is typically available in stores year-round.[50] For example, in the European Union, transport and logistics account for over 9 percent of energy embedded in food consumed across the 27 EU Member States.[51]

Demand for modern bioenergy is growing for use in transport, heating, and electricity. Global production of liquid biofuels has expanded from 16 billion litres in 2000 to more than 100 billion litres in 2011.[52] In addition, the IEA estimates that 2.5–3 billion people rely on biomass and transitional fuels for cooking and heating. Biofuels can provide access to energy for energy-deprived and off-grid communities. When biofuels replace the traditional inefficient combustion of biomass, indoor air pollution and associated poor health are decreased.[53] Finally, biofuels reduce greenhouse gas emissions by replacing fossil fuels in transport and other uses. However, land availability is a limiting factor in biofuel production. Globally, the FAO estimates that growth in biofuel production from 2004 levels to 2030 will require 35 million hectares of land, an area approximately equal to the combined area of France and Spain.[54] In most European countries there are insufficient available land resources to produce the feedstocks required to

Table 2.7 Risks and impacts within the energy-food/food-energy nexus

	Risks	Impacts
Energy-related risks to food security	Dependence on fossil fuel increases the volatility of food prices	Fossil fuel dependency of upstream (e.g. production) and downstream (e.g. transport) for food supply chain
		Price volatility and supply shortages of energy inputs leading to economic and physical risks in the food supply chain
		Social, environmental, and health impacts of traditional biomass cooking methods
	Potential trade-offs between bioenergy production and food crops	Allocation of agricultural products for bioenergy production with possible impacts on food prices
	Risk of energy production on food availability	Possible negative impacts of energy technologies, for example hydropower on food production
Food-related risks to energy security	Overall increase in food production and changing diets raises the energy demand along the food supply chain	Rising demand for energy in agriculture can strain the energy system, especially in areas with the potential to expand irrigated agriculture with pumped water
	Quality and affordability of energy supply can depend on feedstock availability	In energy mixes with bioenergy, quality and affordability of crop-based feedstock can affect energy supply

IRENA. 2015a. Renewable energy in the water, energy and food nexus. Available: http://www.irena.org/DocumentDownloads/Publications/IRENA_Water_Energy_Food_Nexus_2015.pdf comply with blending mandates prescribed in the European Renewables Directive. In Germany, it is projected that by 2030 an estimated 10–11 million hectares of agricultural land would be needed to produce the biomass required to comply with the biofuel blending mandate.[55] As such, biofuels have been criticised for causing food insecurity. Corn, for example, is a major biofuel feedstock in the United States as well as being a staple food crop in many South American and African countries. In 2007, the US Energy Independence and Security Act required the production of 36 billion gallons of biofuels by 2022. This increase in biofuel production led to global food prices increasing by 7 percent.[56]

An overall summary of the risks and impacts of the food-energy/energy-food nexus is summarised in Table 2.7.

2.3   WEF Nexus Pressures

Nexus pressures can result in shortages that put water, energy, and food security for people at risk, hamper economic growth, lead to social and geopolitical tension, and cause irreparable environmental damage: the impacts of WEF nexus pressures are summarised in Table 2.8. As such, identifying the interactions across and between the nexus sectors and improving their efficiency is considered a 'win-win' strategy for human well-being and environmental sustainability for both current and future generations.[57] These interactions are:

- *Interdependencies*: Systems depend on each other; for example, electricity production requires water abstraction and energy is required to extract, treat, and distribute drinking water.
- *Constraints*: Trade-offs between systems; for example, an increased demand in food will increase the amount of water abstracted, reducing the availability of water for other users, for example energy generation.
- *Synergies*: Shared benefits for systems; for example, increasing water and energy efficiency and reducing food waste can reduce WEF nexus pressures overall.[58,59]

Table 2.8 Impact of risks related to WEF nexus pressures

	Direct impacts	Indirect impacts
Impact on governments	Stagnation of economic growth Political unrest Reduced agricultural yields Threats to energy security	Increased social costs from losses of employment and income Potential conflict over scarce natural resources
Impact on society	Increased levels of food insecurity Increased levels of environmental degradation Food and water shortages Social unrest	Migration pressures Irreparably damaged water resources Loss of livelihoods
Impact on business	Export constraints Increased resource prices Commodity price volatility Energy and water restrictions	Lost investment opportunities

2.4 Benefits of Reducing WEF Nexus Pressures

There are many benefits of reducing WEF nexus pressures to achieve green growth that include:

- *Increased resource productivity*: The basic premise of the green economy is the decoupling of resource use and environmental degradation from development (e.g. measured as GDP). Increased resource productivity can be achieved through technological innovation, reducing resource usage, recycling, or reducing waste.
- *Using waste as a resource*: Wastes, residues, and by-products can be turned into resources for other products and services and co-benefits generated. Reducing waste products instead of discharging them into the environment can also reduce environmental degradation and clean-up costs.
- *Stimulating development through economic incentives*: Improvements in resource productivity and resource use generally require investment, for instance research and development. This can be stimulated through

pricing of resources and ecosystem services as well as payments for ecosystem services, which can lead to the private sector acting as the driver of change.

- *Enhancing governance, institutional and policy coherence*: While some new institutions may be required to reduce WEF nexus pressures, taking a nexus approach can strengthen existing institutions so they can build on new links across the nexus sectors while reducing the uncertainty, complexity, and inertia commonly found when integrating a range of sectors and stakeholders.
- *Benefitting from productive ecosystems*: Maintaining and restoring ecosystems through improved management and increased investment can provide multiple services and increase overall environmental, social, and economic benefits.
- *Integrating poverty alleviation and green growth*: The nexus approach supports green growth through smarter use of resources. This strengthens a range of ecosystem services which in turn maintains a healthy environment for humans. The provisioning of clean water and energy improves health and productivity while activities such as green agriculture can generate more rural jobs.
- *Enhancing capacity building and awareness raising*: Enhancing capacity and raising awareness help to deal with the rising complexity of cross-sectoral WEF nexus challenges. Capacity can be increased through monitoring of nexus issues as well as human well-being and equity and developing robust analytical tools that inform policies and regulations across the nexus. Awareness raising can promote sustainable lifestyles and consumption patterns that reduce nexus pressures.[60]

2.5 Realising Reduced WEF Nexus Pressures

A green economy that reduces WEF nexus pressures, which in turn promotes equitable and sustainable growth and ensures a resilient, productive environment can be realised by:

- *Developing an enabling environment to enhance equity in resource access*: As resources become scarce, how resources are allocated, and who will receive these resources will have an impact on social and economic development.

- *Abandoning silo thinking*: Despite strong WEF linkages, practitioners and policymakers continue to approach development programmes and policies in the 'silos' of their respective ministries and government departments. To maximise water, energy, and food security, mechanisms need to be created to raise policymakers' awareness of these issues and promote greater collaboration among government agencies as well as communities, civil society, and the private sector in policy design and implementation.

- *Providing relevant, quantified information and tools across the nexus*: One of the challenges affecting joint planning on reducing nexus pressures is the tendency on many levels to restrict the flow of information as a means of retaining control. However, collaborative approaches required to address WEF nexus pressures must be based on transparency and trust. The sharing and use of observations and information on water, energy, and food is not only essential for planning and implementing activities to reduce resource scarcity but can also shed light on areas where cooperation and resource sharing can result in synergies and more efficient management between the sectors.

- *Developing and disseminating resource-use efficient technologies for enhanced sustainability*: Technology can increase production and conserve scarce resources and be tailored to enhance specific human and environmental conditions. Incentives can be developed to enhance the implementation of new technologies that bring significant benefits to reducing nexus pressures.

- *Optimising market and trade solutions*: Market-based instruments can support resource-use efficiency by changing the behaviour of producers and consumers through market signals. Currently, the pricing of many natural resources, particularly water, is weak due to the complexity and multiple values of the resource across its different uses.[61]

Notes

1. IISD. 2013. The water–energy–food security nexus: Towards a practical planning and decision-support framework for landscape investment and risk management. Available: http://www.iisd.org/pdf/2013/wef_nexus_2013.pdf.

2. Hoff, H. 2011. Understanding the nexus. *Background paper for the Bonn2011 Conference: The water, energy and food security nexus* [Online]. Available: http://wef-conference.gwsp.org/fileadmin/documents_news/understanding_the_nexus.pdf.

3. Houses of Parliament: Parliamentary Office of Science and Technology. 2016. The water-energy-food nexus. *Research Briefing* [Online]. Available: http://researchbriefings.parliament.uk/ResearchBriefing/Summary/POST-PN-0543.

4. IISD. 2013. The water–energy–food security nexus: Towards a practical planning and decision-support framework for landscape investment and risk management. Available: http://www.iisd.org/pdf/2013/wef_nexus_2013.pdf.

5. Global Water Partnership. 2012. Water in the green economy. Available: http://www.gwp.org/Global/ToolBox/Publications/Perspective%20Papers/03%20Water%20in%20the%20Green%20Economy%20(2012).pdf.

6. Brears, R. C. 2016. *Urban Water Security*, Chichester, UK; Hoboken, NJ, John Wiley & Sons.

7. UNEP. 2011a. Biofuels vital graphics: Powering a green economy. Available: http://www.grida.no/publications/vg/biofuels/.

8. Ibid.

9. OECD. 2011b. OECD green growth studies: Energy. Available: https://www.oecd.org/greengrowth/greening-energy/49157219.pdf.

10. IEA. 2017. *What is energy security?* [Online]. Available: http://www.iea.org/topics/energysecurity/subtopics/whatisenergysecurity/.

11. OECD. 2011a. A green growth strategy for food and agriculture. Available: http://www.oecd.org/greengrowth/sustainable-agriculture/48224529.pdf.

12. UNEP. 2011b. Towards a Green Economy: Pathways to Sustainable Development and Poverty Eradication. Available: http://web.unep.org/greeneconomy/resources/green-economy-report.

13. FAO. 1996. The state of food and agriculture. Available: http://www.fao.org/docrep/003/w1358e/w1358e00.htm.

14. UNEP. 2011b. Towards a Green Economy: Pathways to Sustainable Development and Poverty Eradication. Available: http://web.unep.org/greeneconomy/resources/green-economy-report.

15. IEA. 2012b. *Water for energy. Is energy becoming a thirstier resource?* [Online]. Available: http://www.worldenergyoutlook.org/media/weowebsite/2012/WEO_2012_Water_Excerpt.pdf.

16. UNESCO. 2015b. The United Nations World Water Development Report 2015: Water for a sustainable world. Available: https://www.unesco-ihe.org/sites/default/files/wwdr_2015.pdf.

17. IRENA. 2015b. *Renewable energy in the water, energy and food nexus* [Online]. Available: http://www.irena.org/documentdownloads/publications/irena_water_energy_food_nexus_2015.pdf.

18. UNESCO 2012. Managing Water under Uncertainty and Risk.

19. Yang, H., Zhou, Y. & Liu, J. 2009. Land and water requirements of biofuel and implications for food supply and the environment in China. *Energy Policy*, 37, 1876–1885.

20. UNESCO. 2015a. *United Nations World Water Development Report 2015: Water for a sustainable world* [Online]. Available: http://unesdoc.unesco.org/images/0023/002318/231823E.pdf.

21. Ibid.

22. IEA. 2012b. *Water for energy. Is energy becoming a thirstier resource?* [Online]. Available: http://www.worldenergyoutlook.org/media/weowebsite/2012/WEO_2012_Water_Excerpt.pdf.

23. IEA. 2012a. Water for energy. Available: http://www.worldenergyoutlook.org/media/weowebsite/2012/WEO_2012_Water_Excerpt.pdf.

24. Global Water Partnership. 2014. Connecting water and energy. Available: http://www.gwp.org/en/gwp-in-action/News-and-Activities/Connecting-Water-and-Energy/.

25. Siddiqi, A. & Anadon, L. D. 2011. The water–energy nexus in Middle East and North Africa. *Energy Policy*, 39, 4529–4540.

26. IEA. 2012a. Water for energy. Available: http://www.worldenergyoutlook.org/media/weowebsite/2012/WEO_2012_Water_Excerpt.pdf.

27. IEA. 2016. World energy outlook. Available: https://www.iea.org/newsroom/news/2016/november/world-energy-outlook-2016.html.

28. Bobat, A. 2015. Thermal Pollution Caused by Hydropower Plants. *Energy Systems and Management*. Springer.

29. NIWA. 2017. *Sediment* [Online]. Available: https://www.niwa.co.nz/our-science/freshwater/tools/kaitiaki_tools/impacts/sediment.

30. UNESCO 2012. Managing Water under Uncertainty and Risk.

31. FAO. 2014. The water-energy-food nexus: A new approach in support of food security and sustainable agriculture. Available: http://www.fao.org/nr/water/docs/FAO_nexus_concept.pdf.

32. FAO AND WWC. 2015. *Towards a water and food secure future. Critical perspectives for policy-makers* [Online]. Available: http://www.fao.org/3/a-i4560e.pdf.

33. Bhaduri, A., Ringler, C., Dombrowski, I., Mohtar, R. & Scheumann, W. 2015. Sustainability in the water–energy–food nexus. *Water International,* 40, 723–732.
34. Ali, T., Huang, J., Wang, J. & Xie, W. Global footprints of water and land resources through China's food trade. *Global Food Security.*
35. Rasul, G. 2016. Managing the food, water, and energy nexus for achieving the Sustainable Development Goals in South Asia. *Environmental Development,* 18, 14–25.
36. UN-Water. 2015. Wastewater management: A UN-Water analytical brief. Available: http://www.unwater.org/fileadmin/user_upload/unwater_new/docs/UN-Water_Analytical_Brief_Wastewater_Management.pdf.
37. Schuster-Wallace C. J. and Sandford, R. 2015. *Water in the world we want* [Online]. United Nations University Institute for Water, Environment and Health
and United Nations Office for Sustainable Development. Available: http://inweh.unu.edu/wp-content/uploads/2015/02/Water-in-the-World-We-Want.pdf.
38. New South Wales Government. 2009. Irrigation salinity—causes and impacts. Available: http://www.dpi.nsw.gov.au/__data/assets/pdf_file/0018/310365/Irrigation-salinity-causes-and-impacts.pdf.
39. New South Wales Government. 2017. *Irrigation salinity* [Online]. Available: http://www.environment.nsw.gov.au/salinity/solutions/irrigation.htm.
40. Khanom, T. 2016. Effect of salinity on food security in the context of interior coast of Bangladesh. *Ocean & Coastal Management,* 130, 205–212.
41. Carpenter, S. R., Caraco, N. F., Correll, D. L., Howarth, R. W., Sharpley, A. N. and Smith, V. H. 1998. Nonpoint pollution of surface waters with phosphorous and nitrogen. *Ecological Applications,* 8, 559–568.
42. Anderson, D. M., Glibert, P. M. & Burkholder, J. M. 2002. Harmful algal blooms and eutrophication: nutrient sources, composition, and consequences. *Estuaries,* 25, 704–726.
43. Fields, S. 2004. Global Nitrogen: Cycling out of Control. *Environmental Health Perspectives,* 112, A556–A563.
44. FAO. 2014. The water-energy-food nexus: A new approach in support of food security and sustainable agriculture. Available: http://www.fao.org/nr/water/docs/FAO_nexus_concept.pdf.

45. Claudia Hitaj and Shellye Suttles. 2016. Trends in U.S. agriculture's consumption and production of energy: Renewable power, shale energy, and cellulosic biomass. *United States Department of Agriculture Economic Information Bulletin.*

46. Kajenthira Grindle, A., Siddiqi, A. & Anadon, L. D. 2015. Food security amidst water scarcity: Insights on sustainable food production from Saudi Arabia. *Sustainable Production and Consumption, 2*, 67–78.

47. FAO. 2000. The energy and agriculture nexus. Available: http://www.fao.org/docrep/003/x8054e/x8054e00.htm#P-1_0.

48. FAO. 2015. *Fertilizer use to surpass 200 million tonnes in 2018* [Online]. Available: http://www.fao.org/news/story/en/item/277488/icode/.

49. Tallaksen, J., Bauer, F., Hulteberg, C., Reese, M. & Ahlgren, S. 2015. Nitrogen fertilizers manufactured using wind power: greenhouse gas and energy balance of community-scale ammonia production. *Journal of Cleaner Production, 107*, 626–635.

50. Hoolohan, C., Mclachlan, C. & Mander, S. 2016. Trends and drivers of end-use energy demand and the implications for managing energy in food supply chains: Synthesising insights from the social sciences. *Sustainable Production and Consumption, 8*, 1–17.

51. F. Monforti-Ferrario, J.-F. D., I. Pinedo Pascua, V. Motola, M. Banja, N. Scarlat, H. Medarac, L. Castellazzi, N. Labanca, P. Bertoldi, D. Pennington, M. Goralczyk, E. M. Schau, E. Saouter, S. Sala, B. Notarnicola, G. Tassielli, P. Renzulli. 2015. Energy use in the EU food sector: State of play and opportunities for improvement. *JRC Science and Policy Report.* Luxembourg.

52. IRENA. 2015a. Renewable energy in the water, energy and food nexus. Available: http://www.irena.org/DocumentDownloads/Publications/IRENA_Water_Energy_Food_Nexus_2015.pdf.

53. Ben-Iwo, J., Manovic, V. & Longhurst, P. 2016. Biomass resources and biofuels potential for the production of transportation fuels in Nigeria. *Renewable and Sustainable Energy Reviews, 63*, 172–192.

54. UNEP. 2011a. Biofuels vital graphics: Powering a green economy. Available: http://www.grida.no/publications/vg/biofuels/.

55. Ibid.

56. Baier, S. L., Clements, M., Griffiths, C. W. & Ihrig, J. E. 2009. Biofuels impact on crop and food prices: using an interactive spreadsheet. Available: https://www.federalreserve.gov/pubs/ifdp/2009/967/ifdp967.pdf.

57. Ringler, C., Bhaduri, A. & Lawford, R. 2013. The nexus across water, energy, land and food (WELF): potential for improved resource use efficiency? *Current Opinion in Environmental Sustainability*, 5, 617–624.
58. Houses of Parliament: Parliamentary Office of Science and Technology. 2016. The water-energy-food nexus. *Research Briefing* [Online]. Available: http://researchbriefings.parliament.uk/ResearchBriefing/Summary/POST-PN-0543.
59. FAO. 2014. The water-energy-food nexus: A new approach in support of food security and sustainable agriculture. Available: http://www.fao.org/nr/water/docs/FAO_nexus_concept.pdf.
60. Hoff, H. 2011. Understanding the nexus. *Background paper for the Bonn2011 Conference: The water, energy and food security nexus* [Online]. Available: http://wef-conference.gwsp.org/fileadmin/documents_news/understanding_the_nexus.pdf.
61. Ringler, C., Bhaduri, A. & Lawford, R. 2013. The nexus across water, energy, land and food (WELF): potential for improved resource use efficiency? *Current Opinion in Environmental Sustainability*, 5, 617–624.

References

Ali, T., J. Huang, J. Wang, and W. Xie. 2017. Global footprints of water and land resources through China's food trade. *Global Food Security* 12: 139–145.

Anderson, D.M., P.M. Glibert, and J.M. Burkholder. 2002. Harmful algal blooms and eutrophication: Nutrient sources, composition, and consequences. *Estuaries* 25: 704–726.

Baier, S. L., M. Clements, C. W. Griffiths, and J. E. Ihrig 2009. Biofuels impact on crop and food prices: Using an interactive spreadsheet. https://www.federalreserve.gov/pubs/ifdp/2009/967/ifdp967.pdf

Ben-Iwo, J., V. Manovic, and P. Longhurst. 2016. Biomass resources and biofuels potential for the production of transportation fuels in Nigeria. *Renewable and Sustainable Energy Reviews* 63: 172–192.

Bhaduri, A., C. Ringler, I. Dombrowski, R. Mohtar, and W. Scheumann. 2015. Sustainability in the water–energy–food nexus. *Water International* 40: 723–732.

Bobat, A. 2015. *Thermal pollution caused by hydropower plants*. Energy Systems and Management. Cham: Springer.

Brears, R.C. 2016. *Urban water security*. Chichester, UK: John Wiley & Sons.

Carpenter, S.R., N.F. Caraco, D.L. Correll, R.W. Howarth, A.N. Sharpley, and V.H. Smith. 1998. Nonpoint pollution of surface waters with phosphorous and nitrogen. *Ecological Applications* 8: 559–568.

FAO. 1996. The state of food and agriculture. http://www.fao.org/docrep/003/w1358e/w1358e00.htm

———. 2000. The energy and agriculture nexus. http://www.fao.org/docrep/003/x8054e/x8054e00.htm#P-1_0

———. 2014. The water-energy-food nexus: A new approach in support of food security and sustainable agriculture. http://www.fao.org/nr/water/docs/FAO_nexus_concept.pdf

———. 2015. *Fertilizer use to surpass 200 million tonnes in 2018* [Online]. http://www.fao.org/news/story/en/item/277488/icode/

FAO and WWC. 2015. *Towards a water and food secure future. Critical perspectives for policy-makers* [Online]. http://www.fao.org/3/a-i4560e.pdf

Fields, S. 2004. Global nitrogen: Cycling out of control. *Environmental Health Perspectives* 112: A556–A563.

Global Water Partnership. 2012. Water in the green economy. http://www.gwp.org/Global/ToolBox/Publications/Perspective%20Papers/03%20Water%20in%20the%20Green%20Economy%20(2012).pdf

———. 2014. Connecting water and energy. http://www.gwp.org/en/gwp-in-action/News-and-Activities/Connecting-Water-and-Energy/

Hitaj, Claudia, and Shellye Suttles. 2016. Trends in U.S. agriculture's consumption and production of energy: Renewable power, shale energy, and cellulosic biomass. *United States Department of Agriculture Economic Information Bulletin.*

Hoff, H. 2011. Understanding the nexus. *Background Paper for the Bonn2011 conference: The water, energy and food security nexus* [Online]. http://wef-conference.gwsp.org/fileadmin/documents_news/understanding_the_nexus.pdf

Hoolohan, C., C. Mclachlan, and S. Mander. 2016. Trends and drivers of end-use energy demand and the implications for managing energy in food supply chains: Synthesising insights from the social sciences. *Sustainable Production and Consumption* 8: 1–17.

Houses of Parliament: Parliamentary Office of Science and Technology. 2016. The water-energy-food nexus. *Research Briefing* [Online]. http://research-briefings.parliament.uk/ResearchBriefing/Summary/POST-PN-0543

IEA. 2012a. Water for energy. http://www.worldenergyoutlook.org/media/weowebsite/2012/WEO_2012_Water_Excerpt.pdf

————. 2012b. *Water for energy. Is energy becoming a thirstier resource?* [Online]. http://www.worldenergyoutlook.org/media/weowebsite/2012/WEO_2012_Water_Excerpt.pdf

————. 2016. World energy outlook. https://www.iea.org/newsroom/news/2016/november/world-energy-outlook-2016.html

————. 2017. *What is energy security?* [Online]. http://www.iea.org/topics/energysecurity/subtopics/whatisenergysecurity/

IISD. 2013. The water–energy–food security nexus: Towards a practical planning and decision-support framework for landscape investment and risk management. http://www.iisd.org/pdf/2013/wef_nexus_2013.pdf

IRENA. 2015a. Renewable energy in the water, energy and food nexus. http://www.irena.org/DocumentDownloads/Publications/IRENA_Water_Energy_Food_Nexus_2015.pdf

————. 2015b. *Renewable energy in the water, energy and food nexus* [Online]. http://www.irena.org/documentdownloads/publications/irena_water_energy_food_nexus_2015.pdf

Kajenthira Grindle, A., A. Siddiqi, and L.D. Anadon. 2015. Food security amidst water scarcity: Insights on sustainable food production from Saudi Arabia. *Sustainable Production and Consumption* 2: 67–78.

Khanom, T. 2016. Effect of salinity on food security in the context of interior coast of Bangladesh. *Ocean & Coastal Management* 130: 205–212.

Monforti-Ferrario, F., J.-F. Dallemand, I. Pinedo Pascua, V. Motola, M. Banja, N. Scarlat, H. Medarac, et al. 2015. *Energy use in the EU food sector: State of play and opportunities for improvement.* JRC Science and Policy Report. Luxembourg.

New South Wales Government. 2009. Irrigation salinity—Causes and impacts. http://www.dpi.nsw.gov.au/__data/assets/pdf_file/0018/310365/Irrigation-salinity-causes-and-impacts.pdf

————. 2017. *Irrigation salinity* [Online]. http://www.environment.nsw.gov.au/salinity/solutions/irrigation.htm

NIWA. 2017. *Sediment* [Online]. https://www.niwa.co.nz/our-science/freshwater/tools/kaitiaki_tools/impacts/sediment

OECD. 2011a. A green growth strategy for food and agriculture. http://www.oecd.org/greengrowth/sustainable-agriculture/48224529.pdf

————. 2011b. OECD green growth studies: Energy. https://www.oecd.org/greengrowth/greening-energy/49157219.pdf

Rasul, G. 2016. Managing the food, water, and energy nexus for achieving the sustainable development goals in South Asia. *Environmental Development* 18: 14–25.

Ringler, C., A. Bhaduri, and R. Lawford. 2013. The nexus across water, energy, land and food (WELF): Potential for improved resource use efficiency. *Current Opinion in Environmental Sustainability* 5: 617–624.

Schuster-Wallace, C. J., and R. Sandford. 2015. *Water in the world we want* [Online]. United Nations University Institute for Water, Environment and Health and United Nations Office for Sustainable Development. http://inweh.unu.edu/wp-content/uploads/2015/02/Water-in-the-World-We-Want.pdf

Siddiqi, A., and L.D. Anadon. 2011. The water–energy nexus in Middle East and North Africa. *Energy Policy* 39: 4529–4540.

Tallaksen, J., F. Bauer, C. Hulteberg, M. Reese, and S. Ahlgren. 2015. Nitrogen fertilizers manufactured using wind power: Greenhouse gas and energy balance of community-scale ammonia production. *Journal of Cleaner Production* 107: 626–635.

UN-Water. 2013. Water security and the global water agenda.

———. 2015. Wastewater management: A UN-Water analytical brief. http://www.unwater.org/fileadmin/user_upload/unwater_new/docs/UN-Water_Analytical_Brief_Wastewater_Management.pdf

UNEP. 2011a. Biofuels vital graphics: Powering a green economy. http://www.grida.no/publications/vg/biofuels/

———. 2011b. Towards a green economy: Pathways to sustainable development and poverty eradication. http://web.unep.org/greeneconomy/resources/green-economy-report

UNESCO. 2012. Managing water under uncertainty and risk.

———. 2015a. *United Nations World Water Development Report 2015: Water for a sustainable world* [Online]. http://unesdoc.unesco.org/images/0023/002318/231823E.pdf

———. 2015b. The United Nations World Water Development Report 2015: Water for a sustainable world. https://www.unesco-ihe.org/sites/default/files/wwdr_2015.pdf

Yang, H., Y. Zhou, and J. Liu. 2009. Land and water requirements of biofuel and implications for food supply and the environment in China. *Energy Policy* 37: 1876–1885.

3

Policy Tools to Reduce
Water-Energy-Food Nexus Pressures

Introduction

In the transition towards a green economy, a variety of fiscal and non-fiscal policy tools can be implemented to create interdependencies and synergies between the water-energy-food nexus systems while reducing trade-offs between the systems. This chapter will first discuss the various fiscal tools available to reduce nexus pressures in the development of a green economy followed by the various non-fiscal tools available to policy-makers in creating a green economy.

3.1 Fiscal Tools

A green economy seeks to drive green growth, create green jobs, enhance the environment, eradicate poverty, and ensure social equity by shifting the focus of investments from ones that are resource intensive to investments that reduce nexus pressures and create multiple economic, environmental, and social benefits. In this context, fiscal policy tools provide

© The Author(s) 2018
R.C. Brears, *The Green Economy and the Water-Energy-Food Nexus*,
DOI 10.1057/978-1-137-58365-9_3

a critical set of instruments for building the green economy by pricing environmental externalities and redressing social impact.[1,2]

Market-Based Instruments and Pricing

A central aspect of encouraging green growth is integrating the natural asset base into everyday market decisions. This can be achieved through the use of market-based instruments and pricing, including levies, charges, tradeable permits, soft loans, and so forth, which have numerous benefits including:

- *Providing flexibility in achieving natural resource consumption targets*: Economic instruments ensure an overall economy-wide cost of meeting specific targets is reduced by allowing the market to determine how much resource use, or pollution, is achieved.
- *Providing an incentive for the development of new technologies*: Economic instruments provide incentives for firms to develop new technologies that can be sold to others to reduce their resource consumption.
- *Allocating environmental and natural resources to parties who value them the most*: Economic instruments enable the fair allocation of environmental and natural resources and encourage their sustainable utilisation while at the same time raising revenue for governments in the form of resource rents.
- *Guaranteeing self-enforcement by aligning public and private interests*: Economic instruments create a decentralised and self-enforcement system for environmental policies by creating incentives for the proper use of environmental and natural resources, taking away the burden from the government.
- *Helping in cost recovery of publicly provided services*: Economic instruments are applied in the provision of publicly owned or delivered resources, for example, drinking water. Prices are also set at levels that recover the full cost of providing these services. The revenue can then be used to finance the continued provision of these services as well as activities that encourage increased conservation.[3]

Case 3.1 Cape Town's Water Tariffs Based on Level of Scarcity

Because Cape Town is situated in a water-scarce region, the city imposes water restrictions on a permanent basis, with the level of water restrictions dependent on dam storage levels. Cape Town has three levels of water restrictions:

- Level 1 (10 percent water savings): Normally in place
- Level 2 (20 percent water savings): Applicable when dam levels are lower than the norm
- Level 3 (30 percent water savings): Applicable when dam levels are critically low

Since 1 November 2016, Cape Town has been under Level 3 (30 percent water savings), which restricts water usage activities including the prohibition of residents using sprinkler systems, watering their gardens, and washing their cars with hosepipes with municipality-supplied drinking water. To reduce water consumption further, Cape Town will, from 1 December 2016 until further notice, charge all residential, commercial, and industrial water uses Level 3 (30 percent savings) tariffs.[4] For domestic water users, the first 6000 litres remains free, but the next block rates will increase significantly (Table 3.1). Meanwhile, commercial and industrial waters, who are normally charged R 18.77 including VAT per thousand litres (under Level 1 water savings), will now be paying the Level 3 charge of R 25.35 per thousand litres (Table 3.2).[5]

Table 3.1 Cape Town's residential water tariffs

Water 2016–2017 (domestic full) steps	Unit[a]	Level 1 (10% reduction) normal tariffs Rand (incl. VAT)	Level 2 (20% reduction) during level 2 restrictions Rand (incl. VAT)	Level 3 (30% reduction) during level 3 restrictions Rand (incl. VAT)
Step 1 (>0 ≤ 6 kl)	/kl	R 0.00	R 0.00	R 0.00
Step 2 (>6 ≤ 10.5 kl)	/kl	R 14.89	R 15.68	R 16.54
Step 3 (>10.5 ≤ 20 kl)	/kl	R 17.41	R 20.02	R 23.54
Step 4 (>20 ≤ 35 kl)	/kl	R 25.80	R 32.65	R 40.96
Step 5 (>35 ≤ 50 kl)	/kl	R 31.86	R 48.93	R 66.41
Step 6 (>50 kl)	/kl	R 42.03	R 93.39	R 200.16

[a]1kl is a thousand litres

Table 3.2 Cape Town's commercial and industrial water tariffs

Commercial and industrial water use (standard)	Unit	Level 1 (10% reduction) Rand (incl. VAT)	Level 2 (20% reduction) Rand (incl. VAT)	Level 3 (30% reduction) Rand (incl. VAT)
Water	Per kl	R 18.77	R 21.82	R 25.35

Environmental Taxes

Environmental taxes aim to raise the cost of production or consumption of environmentally damaging goods so as to limit their demand.[6] In the green economy, environmental taxes shift the tax burden away from labour—a 'good'—to environmental 'bads' including pollution and the inefficient use of resources.[7] Specifically, by putting a price on environmental externalities, it can reduce pollution and increase resource efficiency in the most cost-effective way and promote behavioural change in consumers and economic sectors.[8,9] Environmental taxes and charges are most commonly levied on energy consumption as well as water.[10,11]

Case 3.2 Alberta's Carbon Levy on Energy

Alberta's carbon levy provides a financial incentive for families, businesses, and communities to lower their emissions by becoming more energy efficient and shifting away from higher emission fuels. As of 1 January 2017, a carbon levy is charged on all fuels that emit greenhouse gases when combusted at a rate of $20/tonne in 2017 and $30/tonne in 2018. The rate is based on the amount of carbon pollution released by the fuel when its combusted, not on the mass of fuel itself. This includes transportation and heating fuels including diesel, gasoline, and natural gas while certain fuels, including marked gas and diesel used on farms and biofuels, are exempt. The levy is in Table 3.3.[12]

Overall, revenue from the carbon levy, along with revenue from the province's Climate Leadership Plan, is expected to raise $9.6 billion, all of which will be reinvested in the local economy, along with rebates to lower-income Albertans. $6.2 billion will be used to diversify Alberta's energy industry and create new jobs with:

- $3.4 billion for large-scale renewable energy, bioenergy, and technology
- $2.2 billion for green infrastructure, for example, public transit
- $645 million for Energy Efficiency Alberta, an initiative to support energy efficiency programmes and services for homes and businesses

Table 3.3 Alberta's carbon levy on major fuels

Type of fuel	1 January 2017	1 January 2018
Marked farm fuels and biofuel	Exempt	Exempt
Diesel	+5.35 ¢/L	+2.68 ¢/L
Gasoline	+4.49 ¢/L	+2.24 ¢/L
Natural gas	+1.011 $/GJ	+0.506 $/GJ
Propane	+3.08 ¢/L	+1.54 ¢/L

Financial Incentives

Many governments use financial incentives to encourage the building of sustainable and resilient infrastructure, promote economic growth hubs, provide services, assist populations to adapt to climate change, and facilitate the adoption of green technologies and practices. This financing mostly comes in the form of subsidies and grants, which are used as 'carrots' to encourage producers as well as consumers to make sustainable consumption choices by closing the price gap for more sustainable products or create significant rebates for their use.[13,14,15,16]

Case 3.3 City of Austin's Bright Green Future School Grants

The City of Austin's Bright Green Future School Grant scheme is a competitive programme that provides funding for school-based sustainability projects up to $3000. The programme, supported by the City of Austin's Office of Sustainability and funded by Austin Resource Recovery, Watershed Protection, the Public Works Department, Austin Transportation, and Austin Energy, is designed to recognise and support innovative projects that will inspire students to become lifelong environmental stewards. Projects that are eligible for grant funding include rainwater harvesting and organic gardening. A key component of being eligible for the grant is that the project actively engages students and members of the community with hands-on involvement and learning.[17]

Payments for Ecosystem Services

Payments for ecosystem services (PES) are one of the principal ways in which a market for ecosystem services can be established with payments made to landowners and others to undertake actions that increase the quantity and quality of desired ecosystem services, which benefit specific or general users, often remotely.[18] Specifically, a PES is a voluntary transaction in which a well-defined ecosystem service or a form of land use likely to secure that ecosystem service is bought by at least one ecosystem service buyer from a minimum of one ecosystem service provider if and only if the provider continues to supply that service. PES are being created to invest in the restoration and maintenance of specific ecosystems and the services they provide. The key characteristic of these PES deals is they focus on maintaining a flow of a specified ecosystem service, for example, clean water or carbon sequestration capabilities in exchange for something of economic value. The critical defining point of a PES transaction is not whether money changes hands and an environmental service is restored or maintained but whether the payment causes the benefit to occur where it would not have otherwise.[19,20] The various types of markets for PES are summarised in Table 3.4.[21]

Table 3.4 Types of markets for payments for ecosystem services

Type of market	Purpose	Description
Public payment schemes for private landowners	To maintain or enhance ecosystem services	These types of PES agreements are country-specific, where governments have established focused programmes. They commonly involve direct payments from a government agency, or another public institution, to landowners and/or managers.
Regulated/ mandated	Have a regulatory cap or floor on the level of ecosystem services to be provided	Established through legislation that creates demand for a particular ecosystem service by setting a 'cap' on the damage to, or investment focused on, an ecosystem. The users of the service, or those responsible for diminishing that service, comply either directly or by trading with others who can meet the regulation at lower cost.
Voluntarily	Companies or organisations engage in a voluntary PES market	Companies or organisations engage in a voluntary market, for example, carbon emissions to enhance their brands, anticipate emerging regulations, or respond to stakeholder and/or shareholder pressure.
Self-organised	Individuals or beneficiaries of ecosystem services contract directly with providers of those services	Voluntary markets are a category of private payments for ecosystem services. Other private PES deals exist where there are no formal regulatory markets and where there is little government involvement. Buyers may be private companies or conservationists who pay landowners to change management practices to improve ecosystem services the buyer wishes to maintain or is dependent on.

Case 3.4 Water Quality Trading in the Ohio River Basin

The Ohio River Basin Water Quality Trading Pilot Project is a first-of-its-kind interstate programme that spans Ohio, Indiana, and Kentucky to evaluate the use of trading by industries, utilities, farmers, and others to meet water quality goals while minimising costs. The water quality trading programme, a market-based approach to achieving water quality goals, allows permitted dischargers to generate or purchase pollution reduction credits from another source. The premise of the water quality trading programme is that:

1. Facility A, for example, a wastewater treatment plant, needs to meet nutrient limits for its water quality permit and therefore water quality trading is one option.
2. To reduce nutrients in the watershed, Facility A pays Farmer B to do a variety of things, for instance, reduce fertiliser use, plant stream side buffers with trees, or keep livestock manure from getting into the waterways, with each conservation practice verified.
3. Nutrient reductions are quantified as credits, for example, equal to one pound of nutrient reduction. Credits are then reviewed and approved by a regulatory agency.
4. Facility A can then use those credits to meet permit requirements.[22]

3.2   Non-fiscal Tools

Governments can use a variety of non-fiscal tools to promote the development of green growth–related technologies and services. They can also use a variety of non-fiscal tools including education, skills development, and awareness-raising to modify the attitudes and behaviour of society towards natural resources to reduce nexus pressures and achieve a green economy.[23,24,25,26,27]

Regulations

Regulations control behaviour and are enforceable through policing institutions and penalties for failure to comply. Regulations encourage or restrict economic activities through the legal system, for example, granting of licences or permits and regulating the labour market. The purpose of regulation is not only to dissuade people from certain behaviours but also to encourage other behaviours. Most regulations work effectively in establishing compliance among affected populations, provided the rules are reasonable and enforced. Regulations work most effectively when the rules are established in consultation with affected populations.[28] Overall, regulations influence green growth by encouraging production efficiency and reducing the number of by-products, while enhancing product market competition. Effective regulations include performance and technology standards, which are useful for reducing negative externalities when market prices fail to reflect some of the cost of economic activities.[29,30]

Case 3.5 Western Australia's Water Efficiency Management Plan

The State of Western Australia requires all businesses using more than 20,000 KL of water a year to complete a Water Efficiency Management Plan (WEMP) to help save water. As part of the programme, large water-using businesses are required to develop a WEMP detailing water-saving targets and actions/initiatives and provide annual progress reports on water-saving targets and actions/initiatives. Businesses are to include in WEMP:

- A profile and description of their business and the current site operations
- Site water use table and usage history
- Water-saving opportunities (including benchmark indicators and targets)
- Water-saving action plan (including time frames)

Once a WEMP is submitted and approved, it is valid for five years. However, if the business changes ownership or the water use increases significantly, a revised WEMP may need to be submitted. Businesses participating in WEMP can be recognised for their outstanding water-saving efforts. Each year, Water Corporation calculates every WEMP participant's water savings in both actual water use and benchmark as a percentage of improvement of the business in water efficiency. Based on the calculated water-saving efforts, businesses can receive one of the awards summarised below:

- Champion: Achieve Gold recognition or better for two consecutive years
- Platinum: More than 50 percent improvement in water efficiency
- Gold: 35–50 percent improvement in water efficiency
- Silver: 25–35 percent improvement in water efficiency
- Bronze: 10–25 percent improvement in water efficiency[31]

Standards and Mandatory Labelling

Mandatory government actions to promote sustainable consumption include performance standards and mandatory labels to limit environmental damage from products when they are consumed or used. These tools are designed to eliminate unsustainable products from the market. The most common sustainability-related performance standards are aimed at reducing energy use, for example, increasing energy efficiency in household appliances.[32]

Case 3.6 Mandatory Energy Efficiency Labelling in Hong Kong

To facilitate the public in choosing energy-efficient appliances and raise public awareness on energy savings, Hong Kong has introduced the Mandatory Energy Efficiency Labelling Scheme (MEELS) through the Energy Efficiency (Labelling of Products) Ordinance. Under MEELS, energy labels are required to be shown on prescribed products for supply in Hong Kong including air conditioners, refrigerating appliances, washing machines, and dehumidifiers. The product's energy label classifies the energy performance of a product type into five grades to help consumers choose energy-efficient products. A product with Grade 1 energy label is among the most energy efficient on the market while a product with Grade 5 is the least efficient.[33,34]

Public Education and Skills Development

Public education can affect green economy innovation in three ways: First, a high level of general and scientific education facilitates the acceptance of technological innovation by society as a whole; second, innovative systems require well-educated researchers and teachers and producers to develop innovations; and third, it is easier for people with higher education and skills to adopt some technological innovations. Continuous skills development (training and retraining) is also essential to improving the matching of skills and demand in sectors that need to adopt productivity and environmentally enhancing technologies and practices.[35]

Case 3.7 Resource-Efficient Wales's Farming Connect Training Programmes

The Welsh Government has set up a number of programmes to help support communities or businesses in becoming more resource efficient. One of the programmes is Farming Connect, which provides farm and forestry businesses in Wales with one-on-one support, advice, guidance, and training in being more efficient in order to safeguard the future of the business. Under this programme, farmers and forestry businesses will be able to identify how to reduce their outputs and improve efficiency. The subsidised service includes training, advice, and mentoring, and is funded up to 80 percent for eligible businesses and is tailored to individual requirements. In addition, the businesses will also receive fully funded access to demonstration farm events, discussion groups, open meetings, business clubs, and workshops and clinics.[36]

Public–Private Partnerships

Investment in physical and knowledge infrastructure is important for overall growth and development. They are critical to the delivery and access to important services and link buyers with sellers and reduce waste, increase productivity, raise profits, and encourage investments in innovative techniques and products.[37] Public–private partnerships (PPPs) are being increasingly recognised as offering feasible solutions to complement or replace public responsibilities for infrastructure and resource-related services. PPPs can be understood as a legal agreement between the public and private sector on the sharing of risks and benefits while embarking on infrastructure development.[38]

Case 3.8 Kuwait's Energy and Water Public–Private Partnership

Kuwait has embarked on a PPP programme that promotes collaboration between the public and private sectors to develop quality infrastructure and services for Kuwaiti citizens. To facilitate PPPs, Kuwait has established the Partnerships Technical Bureau, which aims to utilise private sector skills and expertise in the development of projects in the power and water sectors, among others. One project that is underway is the Independent Water and Power Producer (IWPP) Project. The IWPP Project aims to generate power with a maximum capacity of 2500 MW through a conventional thermal steam power plant. In addition to its power generation capacity the project will include a seawater desalination plant with a capacity of 125 million gallons of water per day. The IWPP Project will be set up as a Special Purpose Vehicle, which will design, build, finance, operate, and maintain the power generation and water production facility for a fixed duration of time. The Special Purpose Vehicle will sign an Energy Conversion and Water and Power Purchase Agreement with the Ministry of Electricity and Water.[39,40]

Stakeholder Participation

Stakeholder participation is a condition that makes sustainable natural resources management and governance more effective. Ideally, decisions are made after all interests have been considered or at least after stakeholders were given the opportunity to express themselves. To make governance more effective, it is important to involve as many stakeholders as possible at various levels. Broad participation offers diverse benefits including: encouraging exchanges and debates on new ideas and information; identifying issues that should be addressed or may have been overlooked; specifying the technical and human capacities necessary to address them; and reaching a consensus on the need for collective action that spurs effective implementation.[41] The overall stakeholder engagement process should include a variety of stakeholders; involve to a greater degree underrepresented affected stakeholders, such as women, youth, or Indigenous populations; be transparent; and have accountability through good communication and information mechanisms.

Case 3.9 Ottawa's Food Policy Council

The Ottawa Food Policy Council, comprised of individuals representing all aspects of the food system, aims to create a food system in Ottawa that emphasises social and economic viability and environmental sustainability through the entire food cycle. The Council's role and scope is to:

- Analyse and monitor policy using a system-wide food lens that takes into account the rural and urban reality of the Ottawa region

- Enable public participation, liaise with community leaders, and consult with relevant stakeholder groups in the policy process
- Cooperate with decision-making bodies in Ottawa including the City of Ottawa, the Ottawa Board of Health, school boards, community organisations, businesses, and others in implementing comprehensive food policies
- Advocate food policy by increasing public awareness through communication with community groups and via formal delegations to decision-making bodies[42]

Information and Awareness-Raising

Information and awareness-raising campaigns are commonly used to promote sustainable consumption of resources, such as the benefits of purchasing environmentally friendly goods and services. As well as campaigns providing information on how sustainable consumption choices can be made, governments can ensure the private sector has access to environmental information to make consumption decisions. A further benefit of easy access to adequate and relevant information is that it can minimise possible conflicts and play a significant role in helping reach a consensus in society.[43,44,45]

Case 3.10 Helsinki's Energy Efficiency Information for Builders

The City of Helsinki Building Control Department advises builders in the permitting phase on how to improve building energy efficiency and how to conduct the energy efficiency survey. The Building Control Department has also been involved in the development of an eco-calculator for buildings which allows builders to assess the ecological qualities of their project including factors such as the: building's potential for heating energy, electricity and water savings; amount of repairable and renewable materials used in the building's construction; amount of energy and emissions used in its construction; and proximity of the building to public transportation services. The eco-calculator will then provide an output of the building compared to current conventional buildings. An example of an eco-house for four people is summarised in Table 3.5.[46,47]

Table 3.5 Helsinki's eco-calculator example

Saving potential		Savings of CO_2 eq/yr	Saving EUR/year
Heating	50–75%	1250–2000	550–820
Domestic hot water	50%	450	200
Household electricity	30%	340	150
Total		2040–2790	900–1170

School Education

School education is one of the most powerful tools for providing individuals with the skills and competencies to become sustainable consumers. Governments can develop good practices in school curricula as well as general consumer education, with the goal of enabling individuals to assess the effects of their own consumption on the environment as well as practice sustainable living.[48]

Case 3.11 United Arab Emirates' 'Our Generation' Initiative

The United Arab Emirates' (UAE) Ministry of Climate Change and Environment, in collaboration with the Ministry of Education and the Environment Agency—Abu Dhabi, has launched the 'Our Generations' initiative to incorporate sustainability and climate change in the school curriculum in support of the national green agenda and strategy. The initiative will be enacted by the 'Sustainable Schools' initiative which was launched by the Environment Agency—Abu Dhabi with the aim of connecting the school community, including students, parents, authority members, and employees, with environmental issues prioritised in the UAE and strengthening their contribution towards achieving the sustainable development goals outlined in the UAE Vision 2021: overall, 148 schools participated in the Sustainable Schools initiative which saw schools reducing environmental footprints, minimising school waste, increasing recycling rates, rationalisation of energy and water consumption, and wastewater reuse. The Our Generations initiative will transfer the experience of the Sustainable Schools initiative to the rest of the schools in the UAE with the aim of imparting awareness on proper environmental practices and healthy lifestyles among the new generation. Overall, the Our Generations initiative will provide students with opportunities to have a strong environmental consciousness and eco-friendly culture as well as participate in practical environmental activities. The initiative will also raise awareness about the fundamental factors that cause environmental problems and equip students with positive attitudes towards their environment.[49,50,51]

Voluntary Labelling

One of the most common tools for influencing sustainable consumer choices is voluntary labelling of products and services. The most viable 'eco-label' schemes are ones whose environmental or social claims are verified by a third party, including governments and non-governmental organisations. These labels can have multi-criteria, which compare products with others in the same category or a number of impacts throughout their lifestyle, or be single social issue labels that refer to a specific environmental or social characteristic of a product.[52,53]

> ### Case 3.12 Germany's Blue Angel Label
>
> In Germany, around 12,000 environmentally friendly products and services from around 1500 companies have been awarded the Blue Angel. The Blue Angel guarantees that a product or service meets high standards with regard to its environmental, health, and performance characteristics. In the process, these products and services are evaluated across their entire lifecycle. Criteria has been developed for each individual product group that must be fulfilled by those products and services awarded with the Blue Label. To reflect technological advances, the Federal Environmental Agency reviews these criteria every three to four years. This ensures that companies constantly improve the environmental performance of their products over time.[54,55]

Clustering Policies

Clusters are geographical concentrations of inter-connected firms and related actors such as specialised service providers, universities, and research centres. There are a variety of terms to describe clusters including 'industrial districts', 'new industrial spaces', and 'knowledge networking'. Clusters enable participants to enjoy a variety of economic benefits including a concentration of specialised suppliers, a large labour market as well as knowledge spillover. Clusters occur at a variety of government levels, from city level right up to supranational, and typically focus on places (leading, lagging, or hub regions), sectors (dynamic, strategically important, socially important) and actors (SMEs, large firms, start-ups, universities, and research centres or combinations of actors). Beyond policies to create the cluster framework governments commonly use a variety of instruments to support the development of clusters including engagement of actors, collective services, and business linkages and collaborative research and development (R&D)/commercialisation. The overall purpose of clusters is to strengthen a particular regional economy.[56,57] Some of the common government instruments used in clusters are summarised in Table 3.6.[58]

Table 3.6 Common government instruments used in clusters

Goal	Action	Instruments
Engage actors	Identify clusters	• Conduct mapping of clusters • Use facilitators and brokers to identify firms that could work together
	Support networks/clusters	• Host awareness-raising events (conferences, cluster education) • Offer financial support for networking activities and events through sponsorship • Benchmark performance • Map cluster relationships
Collective services and businesses linkages	Improve capacity, scale, and skills of suppliers (including SMEs)	• SME business development support • Brokering services and platforms between suppliers and purchasers • Compile general market intelligence • Coordinate purchasing • Establish technical standards
	Increase external linkages (foreign direct investment and exports)	• Market the cluster and region • Assist inwards investment • Partner searches • Supply chain linkage support • Export networks
	Skilled labour force in strategic industries	• Specialised vocational and university training • Support partnerships between groups of firms and educational institutions • Education opportunities to attract students to region

(continued)

Table 3.6 (continued)

Goal	Action	Instruments
Collaborative R&D and commercialisation	Increase links between research and firm needs	• Support joint projects among firms, universities, and research institutes • Co-locate different actors to facilitate interaction • University outreach programmes
	Commercialisation of research	• Ensure appropriate intellectual property framework laws • Technology transfer support services
	Access to finance for spinoffs	• Advisory services for non-ordinary financial operations • Public guarantee and venture capital • Framework conditions supporting private venture capital

Case 3.13 Austria's Ökoenergie-Cluster

In 1999, the regional government of Upper Austria created a network of green energy businesses, the Ökoenergie-Cluster (OEC), to support renewable energy and energy efficiency businesses in the fields of innovation and competitiveness. The cluster partners are companies and organisations in Upper Austria that either manufacture renewable energy and/or energy efficiency technology or are active in the supply chain or as service providers. The main activities of OEC include:

- *Information and communications*: A wide-ranging database of products and services, a website, and newsletters (German/English)
- *Human resource development*: Training, meetings, and workshops
- *Cooperation and technology focus*: Facilitate the development of cooperative projects between cluster partners, universities, research organisations, and other networks/clusters
- *Technology development*: Initiating and supporting research projects of the cluster partners
- *Exports and increasing international focus*: Representing the OEC, supporting businesses in export activities
- *Marketing and PR*: Producing sustainable energy information material, marketing positioning of the OEC, PR work, market research, and development[59]

Resource Mapping

To identify opportunities for targeting resources and policies to meet sustainability goals, governments often need detailed information on the current and future geographical distribution of resource use at various scales. Resource-use maps can contain data on existing and projected resource consumption, present and future population density and land-use type, sources of surplus, large resource consumers, current networks and potential network routes, barriers and opportunities, and socio-economic indicators. These maps can then be used to identify opportunities for investment as well as facilitate stakeholder engagement and raise public awareness on ongoing projects and their benefits.[60]

Case 3.14 Seattle's Get on the Map

The Get on the Map campaign uses an online map of Seattle's neighbourhoods to publicly recognise businesses that are taking actions to save water and energy and reduce waste and pollution. To be put on the map businesses must take a minimum of five green actions including: having a spill kit on hand to prevent spills from entering storm water drains; recycle and compost at the business; purchase and use recycled-content products; assign responsibility for environmental initiatives to a senior manager; measure and report the business' carbon footprint; buy products in bulk; install water-saving fixtures; and any other additional actions. Depending on the number of actions the business takes, the map icon will show up as green, greener, or greenest. The benefits of being on the map include:

- *Gaining recognition*: The Green Business Program will promote the map online, through local media and at community and business events.
- *Share success*: Businesses will receive materials touting their place on the map including website graphics, window clings, and a certificate from the City of Seattle.
- *Receive assistance*: Businesses can receive free one-on-one technical assistance to help them get on the map or darken their shade of green.[61]

Public Procurement

Governments are frequently the largest consumer of goods and services and so they have the power to influence markets towards sustainability through the quantity of their purchases while providing good sustainable consumption examples to their citizens. Often governments have adopted green procurement practices that emphasise the environmental characteristics of products and services.[62]

Case 3.15 Life Cycle Costing in Germany

Under Germany's contract-award law, factoring in life cycle costs into the bid assessment is allowable, and at times mandatory. For example, all federal agencies are required to take life cycle costs into consideration when evaluating bids concerning the procurement of products and services entailing energy consumption. When required, bidders are required to analyse minimised life cycle costs or obtain results using a comparable cost effectiveness evaluation method. To facilitate the ease of doing life cycle analyses, Umweltbundesamt (UBA) provides the UBA Excel Tool, which allows for the evaluation of up to five different modalities and factors in all key cost categories including procurement, operating and disposal costs, and the UBA product-specific Excel tool, which is used to calculate the life cycle costs of computers, computer screens, computer centres, refrigerators, and dishwashers.[63]

Demonstration Projects

Demonstration projects illustrate the feasibility and commercial viability of green economy initiatives and showcase socio-economic benefits to citizens, private building owners, and developers and investors; pilot new policies for uptake by the city council or national government; and build local and institutional capacity and confidence.[64]

Case 3.16 Vancouver's Green and Digital Demonstration Program

Vancouver's Green and Digital Demonstration Program (GDDP) is a joint Vancouver Economic Commission (VEC) and City of Vancouver initiative that provides companies with access to City of Vancouver resources, including buildings, streets, parks, utilities, vehicles, and digital infrastructure, for product testing and showcase opportunities. As part of GDDP, applicants submit proposals to VEC, who then send the short-listed applications to a committee comprised of senior city staff. The range of possible proposals is wide, for example, the use of biofuels in city vehicles and devices that track real-time energy use in city buildings. VEC guides successful candidates through the demonstration process and promotes projects and connects participants to investors, prospective partners, and buyers, while the city is responsible for providing support and staff time to manage the installation and operation of the pilot on available assets as well as provide testimonials/case studies. The overall benefits of participating in GDDP is that companies can refine solutions, accelerate the commercialisation of their technology, attract investment, gain market share, and leverage Vancouver's $31 billion green and innovative brand.[65,66]

Awards and Public Recognition

Role models can be created to strengthen social norms, which are informed rules that are enforced through social ramifications or rewards and therefore guide actions by giving people a general sense of what the majority thinks or does. By creating role models—either winners of competitions or individuals/organisations recognised for their sustainable actions—individuals or communities will likely strive to feel connected with the role model and shift their behaviour to be in accordance with the norm the role model has created.[67,68]

Case 3.17 Sustainable Energy Authority of Ireland's One Good Idea

Ireland's One Good Idea project, run by Sustainable Energy Authority of Ireland, provides an opportunity for students to inspire people to make small lifestyle changes that will use energy more efficiently and address climate change. The project involves teams of two to six students from primary and post-primary schools choosing a topic, researching it before submitting an exciting, creative, and attention-grabbing way of getting the One Good Idea across a variety of audiences including their peers, adults, and the wider community. Both the Top 20 teams from the primary school category and the Top 20 teams from the post-primary school category will receive a workshop with their One Good Idea mentor and activate their campaign. The Top 20 teams from both school categories will also make a presentation pitch explaining their One Good Idea and their awareness campaign. The Top 6 teams from each school category will then attend the National Finals, where they will make their presentation pitch to the panel of expert adjudicators, with the overall winning team for 2016 receiving

€2000 for their school, tablets for each team member, and a €200 voucher for their teacher. The National Finals also had winning categories for Primary Winner 2016, Senior Winner 2016, Junior Winner 2016, and Primary School Runner-Up 2016.[69]

Voluntary Agreements

Voluntary agreements facilitate group actions that aim to improve the productivity and environmental sustainability of a particular economic sector. Voluntary agreements range from initiatives in which participating parties set their own targets, and often conduct their own monitoring and reporting, to initiatives where a contract is made between a private party and a public body, or stakeholder group. By making public their commitments, voluntary agreements are expected to improve the resource efficiency and environmental performance of the sector beyond the level required by existing environmental legislation and regulations. Despite their voluntary nature, the level of enforcement of the agreements can be diverse and the targets set in the agreement can be either general, qualitative goals that seek continuous improvement or specific quantitative targets relative to previous performance such as a reduction of material usage or absolute targets, for example, zero emissions.[70]

Case 3.18 Berlin's Climate Protection Agreements

Berlin's climate change policy is to persuade businesses and organisations operating in Berlin to become active environmentalists, as sustainable climate change mitigation is possible only by harnessing innovative new technology and solutions. As such, Berlin encourages businesses and organisations to take into account the environment and climate in investment decisions and economic activities. Climate Protection Agreements are signed by individual partners such as Berlin's utilities, the housing sector, and various public organisations and are binding obligations. In the Climate Protection Agreements, the signatories pledge to take action in order to achieve the region's climate protection targets including reducing energy consumption, increasing energy efficiency as well as efficiently managing demand for water. These activities are enshrined in comprehensive action plans which are periodically reviewed by the State of Berlin and if necessary revised, ensuring concrete climate protection measures are implemented.[71]

Notes

1. Kai Schlegelmilch and Amani Joas. 2015. Fiscal considerations in the design of green tax reforms. *GGKP Research committee on fiscal instruments* [Online]. Available: http://www.greengrowthknowledge.org/sites/default/files/downloads/resource/Fiscal_Considerations_in_the_Design_of_Green_Tax_Reforms_GGKP.pdf.

2. UNEP. 2013. Green economy fiscal policy briefing paper. Available: http://www.greenfiscalpolicy.org/green-economy-fiscal-policy-briefing-paper-unep-2013/.

3. The Allen Consulting Group. 2006. Market-based approaches to marine environmental regulation. Available: http://www.mfe.govt.nz/sites/default/files/market-based-approaches-marine-regulation-06.pdf.

4. City of Cape Town. 2016b. *Residential water restrictions explained* [Online]. Available: http://www.capetown.gov.za/Family%20and%20home/residential-utility-services/residential-water-and-sanitation-services/2016-residential-water-restrictions-explained.

5. City of Cape Town. 2016a. *Commercial water restrictions explained* [Online]. Available: http://www.capetown.gov.za/Work%20and%20business/Commercial-utility-services/Commercial-water-and-sanitation-services/2016-commercial-water-restrictions-explained.

6. ELD Initiative. 2015. Report for policy and decision makers: Reaping economic and environmental benefits from sustainable land management. Available: http://www.eld-initiative.org/fileadmin/pdf/ELD-pm-report_05_web_300dpi.pdf.

7. EEA. 2014. Resource-efficient green economy and EU policies Available: http://www.eea.europa.eu/publications/resourceefficient-green-economy-and-eu.

8. Ibid.

9. UNCSD Secretariat and UNCTAD. 2011. Trade and green economy. *RIO+20 Issues Briefs* [Online]. Available: https://sustainabledevelopment.un.org/content/documents/132brief1.pdf.

10. OECD. 2008. Promoting sustainable consumption: Good practices in OECD countries. Available: https://www.oecd.org/greengrowth/40317373.pdf.

11. UNDESA. 2012. A guidebook to the green economy. Available: https://sustainabledevelopment.un.org/content/documents/GE%20Guidebook.pdf.

12. Alberta Government. 2017. *Carbon levy and rebates* [Online]. Available: https://www.alberta.ca/climate-carbon-pricing.aspx.

13. K-Water Institute and World Water Council. 2015. Water and green growth: Beyond the theory for sustainable future. Available: http://www.worldwatercouncil.org/fileadmin/world_water_council/documents/mailing/mail_wwf7_documents/Water_and_Green_Growth_vol_1.pdf.

14. UNCSD Secretariat and UNCTAD. 2011. Trade and green economy. *RIO+20 Issues Briefs* [Online]. Available: https://sustainabledevelopment.un.org/content/documents/132brief1.pdf.

15. OECD. 2008. Promoting sustainable consumption: Good practices in OECD countries. Available: https://www.oecd.org/greengrowth/40317373.pdf.

16. UNEP. 2009. Buildings and climate change: Summary for decision-makers. Available: http://www.unep.org/sbci/pdfs/SBCI-BCCSummary.pdf.

17. City of Austin. 2017. *Bright green future school grants* [Online]. Available: http://www.austintexas.gov/brightgreenfuture.

18. Defra. 2010. Payment for ecosystem services: A short introduction. Available: http://www.fwr.org/WQreg/Appendices/payments-ecosystem.pdf.

19. Forest Trends, T. K. G., and UNEP. 2008. Payments for ecosystem services: Getting started. Available: http://www.unep.org/pdf/PaymentsFor EcosystemServices_en.pdf.

20. ELD Initiative. 2015. Report for policy and decision makers: Reaping economic and environmental benefits from sustainable land management. Available: http://www.eld-initiative.org/fileadmin/pdf/ELD-pm-report_05_web_300dpi.pdf.

21. Forest Trends, T. K. G., and UNEP. 2008. Payments for ecosystem services: Getting started. Available: http://www.unep.org/pdf/PaymentsFor EcosystemServices_en.pdf.

22. EPRI. 2015. *The Ohio River Basin water quality trading project* [Online]. Available: http://wqt.epri.com/pdf/EPRI_WQTinfographic.pdf.

23. K-Water Institute and World Water Council. 2015. Water and green growth: Beyond the theory for sustainable future. Available: http://www.worldwatercouncil.org/fileadmin/world_water_council/documents/mailing/mail_wwf7_documents/Water_and_Green_Growth_vol_1.pdf.

24. OECD. 2013. Policy instruments to support green growth in agriculture. *OECD Green Growth Studies* [Online]. Available: http://www.oecd.org/environment/policy-instruments-to-support-green-growth-in-agriculture-9789264203525-en.htm.
25. Brears, R. C. 2016. *Urban Water Security*, Chichester, UK; Hoboken, NJ, John Wiley & Sons.
26. Kai Schlegelmilch and Amani Joas. 2015. Fiscal considerations in the design of green tax reforms. *GGKP Research committee on fiscal instruments* [Online]. Available: http://www.greengrowthknowledge.org/sites/default/files/downloads/resource/Fiscal_Considerations_in_the_Design_of_Green_Tax_Reforms_GGKP.pdf.
27. UNEP. 2013. Green economy fiscal policy briefing paper. Available: http://www.greenfiscalpolicy.org/green-economy-fiscal-policy-briefing-paper-unep-2013/.
28. K-Water Institute and World Water Council. 2015. Water and green growth: Beyond the theory for sustainable future. Available: http://www.worldwatercouncil.org/fileadmin/world_water_council/documents/mailing/mail_wwf7_documents/Water_and_Green_Growth_vol_1.pdf.
29. Ibid.
30. UNCSD Secretariat and UNCTAD. 2011. Trade and green economy. *RIO+20 Issues Briefs* [Online]. Available: https://sustainabledevelopment.un.org/content/documents/132brief1.pdf.
31. Water Corporation. 2017. *Water efficiency management plan programme* [Online]. Available: https://www.watercorporation.com.au/home/business/saving-water/water-efficiency-programs/water-efficiency-management-plan.
32. OECD. 2008. Promoting sustainable consumption: Good practices in OECD countries. Available: https://www.oecd.org/greengrowth/40317373.pdf.
33. Government of Hong Kong Electrical and Mechanical Services Department. 2012. *About mandatory energy efficiency labelling scheme* [Online]. Available: http://www.energylabel.emsd.gov.hk/en/about/background2.html.
34. Government of Hong Kong. 2017. *Mandatory energy efficiency labelling scheme* [Online]. Available: http://www.gov.hk/en/residents/environment/energy/mandatorylabel.htm.

35. OECD. 2015. Analysing policies to improve agricultural productivity growth, sustainably. *A draft framework* [Online]. Available: http://www.oecd.org/tad/agricultural-policies/Analysing-policies-improve-agricultural-productivity-growth-sustainably-december-2014.pdf.

36. Resource Efficient Wales. 2017. *Farming connect* [Online]. Available: http://resourceefficient.gov.wales/programmes/farming-connect/?lang=en.

37. OECD. 2015. Analysing policies to improve agricultural productivity growth, sustainably. *A draft framework* [Online]. Available: http://www.oecd.org/tad/agricultural-policies/Analysing-policies-improve-agricultural-productivity-growth-sustainably-december-2014.pdf.

38. K-Water Institute and World Water Council. 2015. Water and green growth: Beyond the theory for sustainable future. Available: http://www.worldwatercouncil.org/fileadmin/world_water_council/documents/mailing/mail_wwf7_documents/Water_and_Green_Growth_vol_1.pdf.

39. Partnership Technical Bureau. 2017b. *Who we are* [Online]. Available: http://www.ptb.gov.kw/en/Who-We-Are.

40. Partnership Technical Bureau. 2017a. *AIKhairan IWPP (phase I)* [Online]. Available: http://www.ptb.gov.kw/en/AlKhairan-IWPP1.

41. K-Water Institute and World Water Council. 2015. Water and green growth: Beyond the theory for sustainable future. Available: http://www.worldwatercouncil.org/fileadmin/world_water_council/documents/mailing/mail_wwf7_documents/Water_and_Green_Growth_vol_1.pdf.

42. Ottawa Food Policy Council. 2017. *About us* [Online]. Available: http://ofpc-cpao.ca/about-us/.

43. K-Water Institute and World Water Council. 2015. Water and green growth: Beyond the theory for sustainable future. Available: http://www.worldwatercouncil.org/fileadmin/world_wate   r_council/documents/mailing/mail_wwf7_documents/Water_and_Green_Growth_vol_1.pdf.

44. UNDESA. 2012. A guidebook to the green economy. Available: https://sustainabledevelopment.un.org/content/documents/GE%20Guidebook.pdf.

45. OECD. 2008. Promoting sustainable consumption: Good practices in OECD countries. Available: https://www.oecd.org/greengrowth/40317373.pdf.

46. City of Helsinki. 2017. *Energy efficiency information for builders* [Online]. Available: http://www.hel.fi/www/Helsinki/en/housing/construction/efficiency/energy-information/.

47. Rakentajan Ekolaskuri. 2017. *Monta tietä ekologisesti kestävään asumiseen* [Online]. Available: http://www.rakentajanekolaskuri.fi/laskuri.php.

48. OECD. 2008. Promoting sustainable consumption: Good practices in OECD countries. Available: https://www.oecd.org/greengrowth/40317373. pdf.

49. Environmental Center for Arab Towns. 2017. *Sustainable Schools: Environment Agency – Abu Dhabi* [Online]. Available: http://en.envirocitiesmag.com/articles/enviromental-education-and-awareness/sustainable-schools-ead.php.

50. MENA Herald. 2017. *UAE Ministry of Climate Change and Environment launches "Our Generations" initiative* [Online]. Available: https://menaherald.com/en/countries/uae/uae-ministry-climate-change-environment-launches-generations-initiative/.

51. Environment Agency—Abu Dhabi. 2013. *Sustainable Schools (Al Madaris Al Mustadama)* [Online]. Available: https://sustainableschools.ead.ae/en-us/Pages/default.aspx.

52. OECD. 2008. Promoting sustainable consumption: Good practices in OECD countries. Available: https://www.oecd.org/greengrowth/40317373. pdf.

53. ELD Initiative. 2015. Report for policy and decision makers: Reaping economic and environmental benefits from sustainable land management. Available: http://www.eld-initiative.org/fileadmin/pdf/ELD-pm-report_05_web_300dpi.pdf.

54. The Blue Angel. 2017a. *Our label for the environment* [Online]. Available: https://www.blauer-engel.de/en/our-label-environment.

55. The Blue Angel. 2017b. *What is behind it?* [Online]. Available: https://www.blauer-engel.de/en/blue-angel/what-is-behind-it.

56. OECD. 2015. Analysing policies to improve agricultural productivity growth, sustainably. *A draft framework* [Online]. Available: http://www.oecd.org/tad/agricultural-policies/Analysing-policies-improve-agricultural-productivity-growth-sustainably-december-2014.pdf.

57. OECD. 2010. Cluster policies. *OECD Innovation Policy Platform* [Online]. Available: http://www.oecd.org/innovation/policyplatform/48137710.pdf.

58. Ibid.

59. Der Ökoenergiecluster. 2015. The sustainable energy cluster: The network for energy efficiency and renewable energy businesses in Upper Austria. Available: http://www.oec.at/fileadmin/redakteure/ESV/Info_und_Service/Publikationen/Cluster-Profile-fin.pdf.

60. UNEP. 2015. District energy in cities: Unlocking the potential of energy efficiency and renewable energy. Available: http://www.unep.org/energy/portals/50177/DES_District_Energy_Report_full_02_d.pdf.

61. Seattle Public Utilities. 2017. *About the map* [Online]. Available: http://www.seattle.gov/util/ForBusinesses/GreenYourBusiness/GetontheMap/AbouttheMap/index.htm.

62. OECD. 2008. Promoting sustainable consumption: Good practices in OECD countries. Available: https://www.oecd.org/greengrowth/40317373.pdf.

63. The Umweltbundesamt. 2017. *Life cycle costing* [Online]. Available: http://www.umweltbundesamt.de/en/topics/economics-consumption/green-procurement/life-cycle-costing.

64. UNEP. 2015. District energy in cities: Unlocking the potential of energy efficiency and renewable energy. Available: http://www.unep.org/energy/portals/50177/DES_District_Energy_Report_full_02_d.pdf.

65. Vancouver Economic Commission. 2017b. *Green and digital demonstration program* [Online]. Available: http://www.vancouvereconomic.com/gddp/#howitworks.

66. Vancouver Economic Commission. 2017a. *GDDP—How it works* [Online]. Available: http://www.vancouvereconomic.com/gddp/gddp-how-it-works/.

67. Steg, L. & Vlek, C. 2009. Encouraging pro-environmental behaviour: An integrative review and research agenda. *Journal of Environmental Psychology,* 29, 309–317.

68. Barry, M., Ben-Dak, S., Boshoer, V., Capungcol, J., Huang, H., Leer, S., Liu, C., Prince, J. & Yusufova, E. 2014. Encouraging sustainable behavior. Available: http://sustainability.ei.columbia.edu/files/2014/01/Implementation-Plan_FINALlowres.pdf.

69. Sustainable Energy Authority of Ireland. 2017. *About one good idea* [Online]. Available: http://www.seai.ie/Schools/One-Good-Idea/About-one-good-idea/.

70. OECD. 2013. Policy instruments to support green growth in agriculture. *OECD Green Growth Studies* [Online]. Available: http://www.oecd.org/environment/policy-instruments-to-support-green-growth-in-agriculture-9789264203525-en.htm.

71. State of Berlin. 2015. Working together for climate change mitigation in Berlin. Available: http://www.stadtentwicklung.berlin.de/umwelt/klimaschutz/publikationen/download/Klimaschutz-Broschuere_2015_en.pdf.

References

Alberta Government. 2017. *Carbon levy and rebates* [Online]. https://www. alberta.ca/climate-carbon-pricing.aspx

Barry, M., S. Ben-Dak, V. Boshoer, J. Capungcol, H. Huang, S. Leer, C. Liu, J. Prince, and E. Yusufova. 2014. Encouraging sustainable behavior. http://sustainability.ei.columbia.edu/files/2014/01/Implementation-Plan_FINALlowres.pdf

Brears, R.C. 2016. *Urban water security.* Chichester: John Wiley & Sons.

City of Austin. 2017. *Bright green future school grants* [Online]. http://www. austintexas.gov/brightgreenfuture

City of Cape Town. 2016a. *Commercial water restrictions explained* [Online]. http://www.capetown.gov.za/Work%20and%20business/Commercial-utility-services/Commercial-water-and-sanitation-services/2016-commercial-water-restrictions-explained

————. 2016b. *Residential water restrictions explained* [Online]. http://www. capetown.gov.za/Family%20and%20home/residential-utility-services/residential-water-and-sanitation-services/2016-residential-water-restrictions-explained

City of Helsinki. 2017. *Energy efficiency information for builders* [Online]. http://www.hel.fi/www/Helsinki/en/housing/construction/efficiency/energy-information/

Defra. 2010. Payment for ecosystem services: A short introduction. http://www. fwr.org/WQreg/Appendices/payments-ecosystem.pdf

Der Ökoenergiecluster. 2015. The sustainable energy cluster: The network for energy efficiency and renewable energy businesses in Upper Austria. http://www.oec.at/fileadmin/redakteure/ESV/Info_und_Service/Publikationen/Cluster-Profile-fin.pdf

EEA. 2014. Resource-efficient green economy and EU policies. http://www.eea. europa.eu/publications/resourceefficient-green-economy-and-eu

ELD Initiative. 2015. Report for policy and decision makers: Reaping economic and environmental benefits from sustainable land management. http://www. eld-initiative.org/fileadmin/pdf/ELD-pm-report_05_web_300dpi.pdf

Environment Agency—ABU DHABI. 2013. *Sustainable schools (Al Madaris Al Mustadama)* [Online]. https://sustainableschools.ead.ae/en-us/Pages/default. aspx

Environmental Center for Arab Towns. 2017. *Sustainable schools: Environment Agency—Abu Dhabi* [Online]. http://en.envirocitiesmag.com/articles/enviromental-education-and-awareness/sustainable-schools-ead.php

EPRI. 2015. *The Ohio River Basin water quality trading project* [Online]. http://wqt.epri.com/pdf/EPRI_WQTinfographic.pdf

Forest Trends, The Katoomba Group, and UNEP. 2008. Payments for ecosystem services: Getting started. http://www.unep.org/pdf/PaymentsForEcosystemServices_en.pdf

Government of Hong Kong. 2017. *Mandatory energy efficiency labelling scheme* [Online]. http://www.gov.hk/en/residents/environment/energy/mandatory-label.htm

Government of Hong Kong Electrical and Mechanical Services Department. 2012. *About mandatory energy efficiency labelling scheme* [Online]. http://www.energylabel.emsd.gov.hk/en/about/background2.html

K-Water Institute and World Water Council. 2015. Water and green growth: Beyond the theory for sustainable future. http://www.worldwatercouncil.org/fileadmin/world_water_council/documents/mailing/mail_wwf7_documents/Water_and_Green_Growth_vol_1.pdf

MENA Herald. 2017. *UAE Ministry of Climate Change and Environment launches "Our Generations" initiative* [Online]. https://menaherald.com/en/countries/uae/uae-ministry-climate-change-environment-launches-generations-initiative/

OECD. 2008. Promoting sustainable consumption: Good practices in OECD countries. https://www.oecd.org/greengrowth/40317373.pdf

———. 2010. Cluster policies. *OECD Innovation Policy Platform* [Online]. http://www.oecd.org/innovation/policyplatform/48137710.pdf

———. 2013. Policy instruments to support green growth in agriculture. *OECD Green Growth Studies* [Online]. http://www.oecd.org/environment/policy-instruments-to-support-green-growth-in-agriculture-9789264203525-en.htm

———. 2015. Analysing policies to improve agricultural productivity growth, sustainably. *A draft framework* [Online]. http://www.oecd.org/tad/agricultural-policies/Analysing-policies-improve-agricultural-productivity--growth-sustainably-december-2014.pdf

Ottawa Food Policy Council. 2017. *About us* [Online]. http://ofpc-cpao.ca/about-us/

Partnership Technical Bureau. 2017a. *AlKhairan IWPP (Phase I)* [Online]. http://www.ptb.gov.kw/en/AlKhairan-IWPP1

————. 2017b. *Who we are* [Online]. http://www.ptb.gov.kw/en/Who-We-Are

Rakentajan Ekolaskuri. 2017. *Monta tietä ekologisesti kestävään asumiseen* [Online]. http://www.rakentajanekolaskuri.fi/laskuri.php

Resource Efficient Wales. 2017. *Farming connect* [Online]. http://resourceefficient.gov.wales/programmes/farming-connect/?lang=en

Schlegelmilch, Kai, and Amani Joas. 2015. Fiscal considerations in the design of green tax reforms. *GGKP Research Committee on Fiscal Instruments* [Online]. http://www.greengrowthknowledge.org/sites/default/files/downloads/resource/Fiscal_Considerations_in_the_Design_of_Green_Tax_Reforms_GGKP.pdf

Seattle Public Utilities. 2017. *About the map* [Online]. http://www.seattle.gov/util/ForBusinesses/GreenYourBusiness/GetontheMap/AbouttheMap/index.htm

State of Berlin. 2015. Working together for climate change mitigation in Berlin. http://www.stadtentwicklung.berlin.de/umwelt/klimaschutz/publikationen/download/Klimaschutz-Broschuere_2015_en.pdf

Steg, L., and C. Vlek. 2009. Encouraging pro-environmental behaviour: An integrative review and research agenda. *Journal of Environmental Psychology* 29: 309–317.

Sustainable Energy Authority of Ireland. 2017. *About one good idea* [Online]. http://www.seai.ie/Schools/One-Good-Idea/About-one-good-idea/

The Allen Consulting Group. 2006. Market-based approaches to Marine environmental regulation. http://www.mfe.govt.nz/sites/default/files/market-based-approaches-marine-regulation-06.pdf

The Blue Angel. 2017a. *Our label for the environment* [Online]. https://www.blauer-engel.de/en/our-label-environment

————. 2017b. *What is behind it?* [Online]. https://www.blauer-engel.de/en/blue-angel/what-is-behind-it

The Umweltbundesamt. 2017. *Life cycle costing* [Online]. http://www.umwelt-bundesamt.de/en/topics/economics-consumption/green-procurement/life-cycle-costing

UNCSD Secretariat and UNCTAD. 2011. Trade and green economy. *RIO+20 Issues Briefs* [Online]. https://sustainabledevelopment.un.org/content/documents/132brief1.pdf

UNDESA. 2012. A guidebook to the green economy. https://sustainabledevelopment.un.org/content/documents/GE%20Guidebook.pdf

UNEP. 2009. Buildings and climate change: Summary for decision-makers. http://www.unep.org/sbci/pdfs/SBCI-BCCSummary.pdf

———. 2013. Green economy fiscal policy briefing paper. http://www.greenfiscalpolicy.org/green-economy-fiscal-policy-briefing-paper-unep-2013/

———. 2015. District energy in cities: Unlocking the potential of energy efficiency and renewable energy. http://www.unep.org/energy/portals/50177/DES_District_Energy_Report_full_02_d.pdf

Vancouver Economic Commission. 2017a. *GDDP—How it works* [Online]. http://www.vancouvereconomic.com/gddp/gddp-how-it-works/

———. 2017b. *Green and digital demonstration program* [Online]. http://www.vancouvereconomic.com/gddp/#howitworks

Water Corporation. 2017. *Water efficiency management plan program* [Online]. https://www.watercorporation.com.au/home/business/saving-water/water-efficiency-programs/water-efficiency-management-plan

4

The Green Economy and the Water-Energy-Food Nexus in New York City

Introduction

New York City's economy grew at an average rate of 1.7 percent, reaching $670.7 billion in the fourth quarter of 2015. In 2017, the city's annual growth rate is projected to reach 2.7 percent by the fourth quarter.[1] Meanwhile, New York City is home to over 8.5 million people. Over the period 2010–2015, the city's population increased by over 375,000 residents, a growth rate not seen since the 1920s. This is due to a high birth rate, record high life expectancy, and a new influx of people into the city.[2]

4.1 Water-Energy-Food Nexus Pressures

New York City is experiencing a variety of water, energy, and food nexus pressures that are detrimental to the development of a green economy as described below through a variety of examples.

© The Author(s) 2018
R.C. Brears, *The Green Economy and the Water-Energy-Food Nexus*,
DOI 10.1057/978-1-137-58365-9_4

Water

While overall demand for water in New York City has decreased by approximately 30 percent since the 1980s, the energy costs of treating water and wastewater are projected to increase; for instance, the annual variable cost of the wastewater system, including energy and fuel, is expected to increase from $125 million in 2011 to $193 million by 2021, while water system variable costs, including energy and fuel, are projected to increase from $26 million in 2011 to $78 million by 2021. As such, each 1000-gallon reduction in water demand and wastewater flows is estimated to reduce the water and wastewater services variable costs by $0.61 per gallon in 2021. At the system level, a citywide 5 percent water use reduction of over 61 million gallons per day would result in a reduction of water and wastewater treatment costs of $13.63 million by 2021.[3]

Energy

Energy consumption in New York City declined by almost 8 percent from 2005 to 2011, while the city's greenhouse gas (GHG) emissions declined by over 16 percent. Nonetheless, New York City's residential energy prices remain higher than most across the United States. The average price for residential electricity in the city was 29.02 cents per kilowatt-hour (kWh) in 2014, which is nearly 2.4 times greater than the national average of 12.15 cents/kWh. This high price exacerbates the housing cost burden in New York City, where nearly 30 percent of renters allocate more than 50 percent of their household income towards rent each month.[4] Meanwhile, 75 percent of the city's GHG emissions come from in-building activities such as lighting, heating, cooling, and from appliances.[5] Despite the city increasing its share of renewable energy, for example, the city has increased its solar generating capacity from 1 megawatt (MW) in 2006 to 35.7 MW in 2015, about 54 percent of water is used in thermoelectric plants.[6,7]

Food

New York City's food distribution system is one of the largest in the United States. It serves 8.4 million residents, over 60 million annual tourists, and hundreds of thousands of commuters on a daily basis.[8] The city's agencies alone serve more than 245 million meals and snacks per year. In addition to the over 171 million meals and snacks served in schools, New York City, via directly or through its non-profit partners, serves an additional 74 million meals in homeless shelters, child care centres, after-school programmes, public hospitals, and so forth.[9] The demand for food will likely increase with the city's expected exceedable population of 9.5 million by 2020. The result is total inbound food is projected to rise from 33.5 million tonnes in 2002 to 54 million tonnes in 2035, a 61 percent increase.[10] However, already a significant portion of the population experiences food insecurity with more than 1.3 million New York City residents, or 16.4 percent, being food insecure, resulting in these residents falling short of an adequate diet by 242 million meals in a single year.[11]

4.2   Water: Fiscal Tools

New York City has implemented a variety of water-related fiscal tools that create interdependencies and synergies between the nexus systems while reducing trade-offs between the systems in the development of a green economy.

Residential Water Efficiency Programme: Retrofitting Toilets

To reduce indoor residential water use the city's Department of Environmental Protection (DEP) initiated in 2014 a toilet replacement programme. To service the programme DEP has contracted it out to wholesale plumbing supply vendors throughout the city to accept customer vouchers, provide adequate plumbing fixtures, and track sales. To receive a high-efficiency toilet, customers log on to a DEP-designed web portal to receive a $125 voucher per toilet. Applicants then choose from a range of approved high-efficiency toilets. Once the toilet has been selected, the wholesale plumber contractors have 90 days to install the product.[12]

4.3   Water: Non-fiscal Tools

New York City has implemented a variety of water-related non-fiscal tools that create interdependencies and synergies between the nexus systems while reducing trade-offs between the systems in the development of a green economy.

NYC DEP's Annual Water Art and Poetry Contest

In 2015, New York City's DEP held its 29th Annual Water Resources Art and Poetry Contest. Around 1350 students between grade 2 and 12 from New York City and watershed communities created more than 1400 original pieces of artwork and poetry that show appreciation for New York City's shared water resources and the importance of water conservation. In addition, through art and poetry students expressed their understanding of the city's water supply and wastewater treatment systems. All participating students were honoured as DEP Water Ambassadors and received a certificate recognising their artistic and poetic contribution. In addition, a panel of judges selected 60 students as DEP Water Champions. Winning entries were selected on originality, artistic ability, and understanding of one or more contest themes. The themes were: Water—a precious resource; The New York City water supply system; The New York City wastewater treatment system; Climate change; and Water stewardship. Overall, the programme raises awareness about the importance of clean, high-quality drinking water and the resources required to maintain the city's water and wastewater treatment systems.[13]

Mayor's Water Challenge to Hotels

In 2013, DEP partnered with the Hotel Association of NYC Inc. and the Mayor's Office to develop the Mayor's Water Challenge to hotels: a public–private partnership designed to encourage hotels to reduce their annual water consumption by 5 percent: The challenge is voluntary and is designed to encourage participants to match the city's commitment of reducing citywide water consumption by 5 percent by 2020.

The Mayor's Water Challenge is for a 12-month period in which DEP tracks, monitors, and reports changes in water consumption attributable to the implementation of new water conservation measures. Challenge participants are asked to submit a formal water conservation plan which summarises the facility's plan to achieve a targeted 5 percent reduction in water use from the baseline water year: Prior to participation, each hotel will have an automated meter reader installed to track water consumption and establish a baseline profile of water use.[14]

During the programme, participants will attend regular meetings with other Water Challenge participants, attend annual one-on-one meetings with the Mayor's office and DEP staff, and convene voluntary working groups with DEP and the Mayor's office. The city will also facilitate regular meetings with industry experts and organisations to provide contacts, best practices, and technical assistance needed to realise significant water savings. In addition, participants will receive a monthly summary progress and benchmarking report. DEP will compile monthly data for all participating hotels and distribute a personalised report to hotel managers. This report shows current monthly consumption levels for each hotel as well as their water use intensity (gallons per square feet), benchmarked against other participating hotels. The reports are customised so participants can only view data belonging to their specific facility but can see anonymous comparisons of their hotel with other hotels in the programme.[15]

Mayor's Water Challenge to Restaurants

Following the completion of the Mayor's Water Challenge to hotels, DEP, in 2014, partnered with the US EPA, Con Edison, the New York

City chapter of the New York State Restaurant Association, the New York State Energy Research and Development Authority, and Alliance for Water Efficiency to develop the NYC Water Challenge to restaurants. The public–private partnership challenges a select group of 30 New York City restaurants to reduce their annual water consumption by 5 percent. NYC DEP and participating restaurants will work together to identify and realise water savings. There are four components of the water challenge: developing a water conservation plan, four conservation workshops, monthly reports, and getting recognition of achievement through press releases. The participating restaurants have already established their baseline water consumption trends with many also having analysed their energy consumption. To reduce water, and energy, consumption, the restaurants are auditing their kitchens for water and energy use, re-designing or upgrading kitchens to include the re-use of greywater, or capturing of rainwater for irrigation. At the end of the 12-month period participating restaurants that reduce their water consumption by 5 percent or more will be recognised at an award ceremony. The results and lessons learnt from the challenge will be published in a booklet titled *NYC DEP's Restaurant Managers Guide to Water Efficiency*. This booklet will be available as part of DEP's non-residential educational resources.[16]

Monitoring Water Consumption

DEP is working to give customers more information on their water consumption. Providing consumption information on a timely basis empowers customers to detect inefficiencies including leaks via the My DEP Account web portal. Currently more than 324,000 customers have signed up for My DEP Account where customers can view their water usage, bills, and payment history online. Small customers can view their meter readings four times a day while larger customers can view their meter readings on an hourly basis. This enables all types of customers to monitor their consumption and be more aware of consumption patterns. DEP has also included in My DEP Account an option that allows customers to receive a leak alert if their consumption triples for five consecutive days. This alert helps customers quickly detect leaks and fix them, saving water and money. Over 220,000 customers have signed up for the leak alerts while large customers can customise their leak detection parameters.[17]

DEP Watershed Agricultural Programme

Most of New York City's drinking water comes from reservoirs adjacent to productive farmland. To preserve the quality of its source water NYC's DEP, in partnership with the Watershed Agricultural Council, has initiated the Watershed Agricultural Programme to promote best agricultural practices to prevent harmful runoff into the water supply and help them maintain financial sustainability. The programme includes the Nutrient Management Programme, farmer education, and Pure Catskills marketplace.

Nutrient Management Programme

The Nutrient Management Programme is a multi-agency initiative that assists farmers in improving phosphorus and pathogen management through manure, fertiliser, and animal feed management. The programme provides nutrient management credits to encourage good stewardship of manure resources to improve water quality, providing the Watershed Agricultural Programme the means to enhance participation by farmers in the Nutrient Management Programme. The programme also offers its staff to help dairy and beef farmers implement better feed nutrient management practices.

Farmer Education Programme

The Farmer Education Programme aims to improve water quality and farm viability by providing education programmes that enhance farmers' abilities to manage their operations more profitably and in a way that protects their natural resources. Currently the programme focuses on hands-on training and practical tools for profitable education. Farmers are taught about new technology and new crops and markets for both new and established farmers.

Pure Catskills Marketplace

Pure Catskills is a regional, buy-local-products campaign developed to improve the economic viability of the local community and preserve water quality in the NYC watershed region. Currently, Pure Catskills represents nearly 300 farmers and forest-based businesses, restaurants, local artisans, accommodations, and non-profit organisations. As part of Pure Catskills the online Pure Catskills marketplace was launched to sell high-quality and authentic farm, food, and forest products grown, raised, and made throughout the region. By purchasing products via the marketplace consumers are investing in healthy forests, farmland protection, and clean drinking water.[18]

4.4   Energy: Fiscal Tools

New York City has implemented a variety of energy-related fiscal tools that create interdependencies and synergies between the nexus systems while reducing trade-offs between the systems in the development of a green economy.

Solar Panel Tax Abatement

In 2008, New York City passed legislation to provide a four-year tax abatement, or tax relief, of 5–8.75 percent of solar panel–related expenditures, up to $62,500 or the building's tax liability, whichever is less.[19]

4.5   Energy: Non-fiscal Tools

New York City has implemented a variety of energy-related non-fiscal tools that create interdependencies and synergies between the nexus systems while reducing trade-offs between the systems in the development of a green economy.

Greener, Greater Buildings Plan

In 2009, New York City enacted the Greener, Greater Buildings Plan (GGBP) to increase energy efficiency and reduce GHG emissions of large existing buildings. GGBP consists of a range of local laws that are designed to provide information about buildings' energy use and requires the implementation of some cost-effective upgrades, which include the following:

Local Law 84: Benchmarking

Local Law 84 requires private sector owners of buildings greater than 50,000 gross square feet or located on a lot with more than 100,000 gross square feet of space; multiple private sector properties totalling over 100,000 square feet; and public sector buildings over 10,000 square feet to annually measure their energy and water consumption in a process called benchmarking.[20] Local Law 84 standardises this process by requiring building owners to submit this information to the NYC Department of Buildings annually using a free online tool provided by the US EPA called ENERGY STAR Portfolio Manager. The information provided enables building owners to compare their energy and water consumption with similar buildings and tracks progress year to year to help in energy and water efficiency planning. In addition, the information is publicly disclosed online to provide transparency about the energy and water use of the city's largest buildings.[21]

Local Law 85: NYC Energy Conservation Code

This law requires all New York City buildings that undergo any renovation or alteration to meet the most current state energy code. Prior to 2009, buildings were only required to meet the energy code if renovations altered more than half the building's system. Closing this loophole now means buildings can accrue energy savings following incremental upgrades.[22]

Local Law 87: Energy Audits and Retro-commissioning

Local Law 87 requires that all buildings greater than 50,000 gross square feet must undergo energy audits and retro-commissioning measures every ten years where retro-commissioning is the testing and fine-tuning of existing building systems to confirm they are operating as designed and as efficiently as possible.[23]

Local Law 88: Lighting Upgrades and Sub-metering

Local Law 88 requires lighting upgrades and energy use sub-metering in all commercial buildings by 2025, allowing for upgrades to be made when leases turn over.[24]

NYC Carbon Challenge: Following the City's Lead

To reduce emissions, New York City has enacted a comprehensive energy management strategy across the city's portfolio of around 4000 buildings. This includes conducting energy audits, performing retro-commissioning studies, and undertaking energy-efficient upgrades. In addition, the city has invested in renewable energy sources including solar power, piloted leading technologies across its portfolio and trained nearly 2000 building owners in efficient operations.[25]

NYC Carbon Challenge

To encourage the private sector and institutional sector to follow the city's lead in reducing GHG emissions the voluntary NYC Carbon challenge was also established. Over 50 participants including universities, hospitals, commercial firms, multifamily buildings including residential cooperatives, condominiums, rental buildings as well as Broadway theatres have signed up to the challenge of reducing their emissions by 30 percent or more in ten years.[26]

NYC Retrofit Accelerator

To assist building owners, the city launched the NYC Retrofit Accelerator in 2015 to provide free technical assistance and advisory services for building owners to 'go green' through energy efficiency, water conservation, and clean energy upgrades.[27] The retrofit accelerator will include a team of customer service and building experts to provide technical assistance to help remove the numerous complexities of undertaking projects. In particular, the retrofit accelerator team can help decision-makers select contractors, explain the necessary permits, and navigate financial and incentive programmes to help cover the costs. Building maintenance staff will also be given access to training and education to help improve the quality of operations and maintenance. A key aspect of the retrofit accelerator is that it will seek to complement, not replicate, existing financing, incentive, and assistance programmes available in New York City.[28,29,30]

Innovative Demonstrations for Energy Adaptability Programme

In 2014, the Department of Citywide Administrative Services launched the Innovative Demonstrations for Energy Adaptability (IDEA) programme to identify emerging energy technologies and evaluate their potential for installation across the city's building portfolio. IDEA engages vendors of emerging and underutilised energy technologies to test their solutions in city buildings. The pilot, focusing on building controls, led to agreements with 12 vendors for demonstrations of new technologies in 23 buildings across eight city agencies. As IDEA expands new phases will target specific topics identified by city agencies. Demonstrations will be prioritised on their potential for energy cost savings, GHG emission reductions, resiliency benefits, and job creation potential. Following completion, the Department of Citywide Administrative Services will evaluate performance data from demonstrations and choose solutions for replication in other city buildings, which could lead to larger-scale procurements. The Department of Citywide Administrative Services will then share success stories with the private sector through a central clearinghouse of case studies, performance data, and analysis.[31]

NYC Solar Map

NYC Solar partnership has developed NYC Solar Map, which is a free, interactive tool available to help all New Yorkers to estimate their buildings' solar potential, savings, and pay-back period taking into account financial incentive programmes in the city.[32]

NYC Multifamily Solar Guide

NYC Solar partnership has published the New York City Multifamily Solar Guide for residents, board members, property managers, and other multifamily building stakeholders who are interested in pursuing solar energy so they can understand the benefits of solar, what is required to install solar, and the steps for moving forward with solar projects.[33]

NYC Energy & Water Performance Map

New York City's Energy & Water Performance Map shows New Yorkers how their buildings compare to other similar buildings in terms of energy and water management. This empowers all New Yorkers to understand the energy, water, and climate change impacts of building spaces they rent or buy and to identify top performing buildings in the city. The map's interface provides energy and water efficiency details for specific buildings and how their performance compares with other buildings in the city. Data for the map comes from NYC's Local Law 84, which requires private buildings over 50,000 square feet and public sector buildings over 10,000 square feet to report their energy and water consumption each year for public disclosure. Overall the performance map enables users to compare energy and water consumption data for over 2.3 billion square feet in 23,000 private sector buildings and 281 million square feet in over 3000 public sector buildings.[34, 35]

4.6   Food: Fiscal Tools

New York City has implemented a variety of food-related fiscal tools that create interdependencies and synergies between the nexus systems while reducing trade-offs between the systems in the development of a green economy.

Health Bucks

$2 coupons are redeemable for fresh fruits and vegetables at all New York City farmers' markets as part of New York City's Health Bucks programme. The programme enables low-income New Yorkers to buy fresh, locally grown produce at more than 140 farmers' markets across the city. Health Bucks are distributed as a Supplemental Nutrition Assistance Program (SNAP) incentive: For every $5 spent in SNAP benefits (food stamps) at the market, beneficiaries receive a $2 Health Bucks coupon.[36]

4.7   Food: Non-fiscal Tools

New York City has implemented a variety of food-related non-fiscal tools that create interdependencies and synergies between the nexus systems while reducing trade-offs between the systems in the development of a green economy.

Food Waste Challenge

In 2012, NYC launched the Mayor's Food Waste Challenge, a voluntary programme encouraging city restaurants to commit to help reducing land-filled food waste by committing to a 50 percent food waste diversion goal. Specifically, participants were challenged to reduce the amount of their food waste that goes to landfill by 50 percent from their base year, of which no more than 10 percent can come from non-organic waste recycling and reduction. As part of the challenge the programme required participants to conduct a baseline waste generation to determine their volume or weight of the total waste generated and the composition and monitor changes on at least a quarterly basis. To successfully complete the challenge each participant had to meet the 50 percent food waste diversion goal subject to verification by Office of Long-Term Planning and Sustainability (OLTPS) or a neutral third party. The programme consisted of regular partner meetings; workshops on topics including simple ways to measure waste generation and effective messaging and training; access to resources and technical expertise; and promotion and recognition including official participant listing on the Mayor's Food Waste Challenge website, inclusion of company name or logo, successful completion profiles, the ability to self-promote participation in the challenge including the use of the Mayor's Carbon Challenge logo on company documents and social media and placement of a Food Waste Challenge participant sign in the restaurant window.[37]

Sustainable Food Procurement: School Food

New York City's Department of Education (DOE) has the largest school food service programme in the United States, with around $200 million in annual food purchases, and uses its purchasing power to lead the market. DOE's Office of SchoolFood (SchoolFood) has made the procurement of local and fresh food a priority as well as sourcing sustainable and healthy products. Actions to date include:

- Replacing all Styrofoam trays with compostable plates made from 100 percent renewable materials, removing 128 million polystyrene trays from the city's landfills and waterways.
- SchoolFood along with five other school districts that form the Urban School Food Alliance announced a new antibiotic-free standard for chicken supplied to schools.
- Educating students about the food offered to them by launching a student-driven visual menu. Students from across New York City visit SchoolFood to evaluate new products and learn about the school food service.
- Sourcing produce, milk, and yoghurt that was locally or regionally grown or produced.[38]

Garden to Café Programme

With support from the SchoolFood the Garden to Café programme connects school gardens with their cafeterias with the goal of increasing students' knowledge of healthy food, farming, and the local food system. The programme features harvest events where food grown in the school garden is incorporated into school meals via menu items or the cafeteria for tasting. To date over 120 schools have registered with the programme.[39]

NYC School Gardens Initiative

Grow to Learn NYC, an initiative of the non-profit GrowNYC, in partnership with the NYC Park's GreenThumb Programme, which supports community gardens, and DOE's SchoolFood programme, which connects school gardens with the cafeteria through seasonal harvests and educational activities, has launched the NYC School Gardens initiative to inspire, facilitate, and promote the creation of a school garden in every public school in New York City. Schools work directly with Grow to Learn NYC to ensure their garden programmes are sustainable and transformative for student learning in the cafeteria and classroom. As part of the initiative Grow to Learn NYC provides:

- Funding twice a year in the form of mini-grants of $500–$2000 for materials
- Technical help through skill-building workshops, harvest events, garden builds, one-on-one support
- A citywide network that enables school gardeners to share knowledge and resources
- Advocacy to build citywide support and value for school gardens[40]

New York City Housing Authority's Garden and Greening Programme

The New York City Housing Authority's (NYCHA) Garden and Greening Programme supports over 700 gardens (around half of which are food producing) and three urban farms managed by local community partners. The programme is guided by the Next Generation NYCHA strategic plan that aims to increase sustainability, reduce the carbon footprint of NYCHA, and increase workforce opportunities for NYCHA residents. Moving forward new farms under the programme are expected to produce approximately four tonnes of fresh produce and engage around 500 community residents each year while offering a venue for training and service for young NYCHA residents in collaboration with the organisation Green City Force.[41]

Food Sector Job Training Programmes

The Department of Small Business Services (SBS) provides training grants to small business owners with the goal of helping New York City businesses and their employees succeed. Agricultural and food-related businesses are eligible for support including the NYC Business Solutions' Customised Training Programme.

NYC Organics Collection: Voluntary Residential Organic Waste Collection

In 2013, New York City Council passed Local Law 77 requiring the New York City Department of Sanitation (DSNY) to implement a voluntary residential organic waste kerbside collection pilot programme and a school organic waste collection pilot programme. NYC Organics Collection operates a similar strategy to the current NYC recycling programme, which collects recyclable materials from the kerb on specific days of the week. By 2015, the programme has collected organic waste from over 100,000 households, 700 schools, and agencies and institutes across the city. By 2018, NYC aims to expand the Organics Collection programme to serve all New Yorkers through kerbside collection or convenient local drop-off sites.[42]

Organic Waste to Energy Pilot Programme

In 2013, NYC DEP launched a pilot programme at the Newtown Creek Wastewater Treatment Plant to process food waste in anaerobic digesters, increasing the production of renewable biogas on-site. Under the pilot programme Waste Management takes in food waste from the company's transfer station where it is grounded up into slurry and transported in sealed tankers to the Newtown Creek plant. At the plant the food waste slurry is added to the plant's existing anaerobic digesters. In the future this gas will be refined into pipeline quality renewable natural gas and pumped into the local National Grid gas system.

Organic Waste to Energy Demonstration Project

In 2015, NYC DEP expanded the pilot programme into a three-year demonstration project in which pure food waste from restaurants and other food service establishments will be added to the plant's feedstock. The plant will initially accept 50 tonnes of food waste slurry per day, increasing to 250 tonnes per day by 2018. If the demonstration project is successful the plant will be able to process up to 500 tonnes of food waste per day to produce enough energy to heat over 5000 homes.[43]

Zero Waste Commercial Organics Rules

In New York City, all large-scale commercial food establishments must start separating organic waste from 19 July 2016 onwards. Specifically, the establishments covered by the Commercial Organics Rules are:

- All food service establishments in hotels with 150 or more rooms
- All food service vendors in arenas and stadiums with seating capacity of more than 15,000 people
- Food manufacturers with a floor area of at least 25,000 square feet
- Food wholesalers with a floor area of at least 20,000 square feet

Businesses covered by this regulation are given the option to arrange for collection by a private firm, transport organic waste themselves, or process the material on-site with suitable processing methods including composting and aerobic or anaerobic digestion.[44]

Zero Waste Challenge

The Zero Waste Challenge, which began in mid-February and concluded in mid-June 2016, was initiated to help businesses that are affected by these new rules increase their recycling and organics separation and diversion efforts by challenging them to divert as much material from landfill and incineration as possible. Participants were provided with training materials, a customised monthly waste tracking report, case studies provided by experts, and the opportunity to attend monthly workshops led by the Mayor's Office and featured guest speakers in the areas of waste reduction and diversion. Participants of the challenge were recognised at the end for:

- Diverting 50 percent of waste from landfill and incineration
- Diverting 75 percent of waste from landfill and incineration
- Reaching 90 percent of waste diverted from landfill and incineration
- Greatest overall waste diversion rate from landfill and incineration
- Greatest amount of food donated to local charities and organisations to feed hungry New Yorkers
- Greatest overall waste diversion rate from landfill and incineration by category of business
- Most successful or innovative source reduction effort[45]

4.8   Case Study Summary

New York City has implemented a variety of initiatives to create interdependencies and synergies between the nexus systems while reducing trade-offs between the systems in the development of a green economy.

Water

- To reduce water consumption in the city, the DEP has implemented a toilet retrofit voucher programme.
- To raise awareness on the need to conserve water, DEP has an annual art and poetry contest for students with different water-related themes.

- To encourage businesses to reduce water usage, the city challenges specific water-intensive sectors to reduce their consumption with the city providing skills and training for participating businesses.
- To better manage water resources, DEP is enabling customers to view their water consumption data daily, enabling the faster detection of leaks.
- To protect the city's water source from agricultural practices, the city partners with agricultural stakeholders to reduce agricultural runoff and protect waterways. At the same time the partnership promotes local food produce grown in a responsible manner that protects drinking water.

Energy

- To encourage the installation of solar photovoltaic (PV) systems, the city offers a tax rebate.
- To reduce energy use and increase energy efficiency in buildings, the city has enacted a series of local laws including the mandatory benchmarking of large private sector buildings that also enables building owners to compare their energy and water consumption with similar buildings and track their progress. The city also requires large buildings to complete energy audits and fine-tune energy systems in existing buildings every decade.
- New York City has challenged the private sector to reduce GHG emissions by investing in renewable energy systems and improving the energy efficiency of large privately owned buildings.
- To assist private building owners in reducing GHG emissions further the city provides free technical assistance and advisory services for building owners to implement energy and water efficiency initiatives and clean energy upgrades.
- To encourage the development of new technologies the city enables vendors to test energy-efficient technologies in city-owned buildings with the city potentially offering procurement contracts in the future.
- The city has developed an Energy & Water Performance Map, which shows New Yorkers how their buildings compare to other similar buildings, helping residents identify top performing buildings in the city.

Food

- The city provides vouchers for low-income residents to purchase locally grown produce at farmers' markets across the city.
- The city has challenged restaurants to reduce food waste with the city providing free advice and training for participants, along with public recognition of restaurants.
- The DOE has made the procurement of locally produced healthy food a priority that includes linking school gardens with their cafeteria to increase students' knowledge of healthy local and food production. To increase the participation of schools, the city has partnered with a variety of stakeholders to implement gardens in every public school throughout the city.
- The city has a programme that supports community gardens and urban farming initiatives throughout the city including on NYCHA property. By developing gardens at housing authority properties, it not only increases access to healthy food but also enables the city to offer training programmes for young residents as well as working opportunities for residents.
- The city has initiated a voluntary residential organic waste collection pilot programme as well as a school organic waste collection pilot programme to reduce food waste entering local landfills. The city has also launched an organic waste to energy pilot programme that involves taking in pure waste from restaurants and other food service establishments to generate energy for heating homes.
- New York City has initiated a zero waste commercial organics rule that all large-scale commercial food establishments must start separating organic waste from mid-2016 onwards. To help businesses affected by these new rules, the city launched a Zero Waste Challenge that involved setting organic waste diversion targets and the city providing various types of assistance including customised tracking reports, case studies, monthly meetings, and workshops.

Overall, New York City uses a variety of fiscal and non-fiscal tools to create interdependencies and synergies between the nexus systems while reducing trade-offs between the systems in the development of a green economy. These tools are summarised in Table 4.1.

Table 4.1 New York City case summary

Tool	Tool type	Policy title	Description	WEF sectors addressed
Fiscal	Market-based instruments and pricing	Pure Catskills	Is a regional, buy-local-products campaign developed to improve the economic viability of the local community and preserve water quality in the NYC watershed region. By purchasing products via the marketplace consumers are investing in clean drinking water.	Food, water
	Environmental taxes	Solar panel tax abatement	A property tax abatement is available for the installation of solar PV systems.	Energy
	Financial incentives	Residential water efficiency	Residents receive a voucher for the retrofitting of old toilets.	Water
		Health Bucks	The coupons help low-income New Yorkers to buy fresh, locally grown produce.	Food

(continued)

Table 4.1 (continued)

Tool	Tool type	Policy title	Description	WEF sectors addressed
Non-fiscal	Regulations	Benchmarking energy and water usage	Private buildings of certain sizes are required to measure their energy and water usage and report it to the city enabling comparisons to be made between buildings.	Energy, water
		NYC Energy Conservation Code	All buildings being renovated must meet the current state energy code.	Energy
		Energy audits and retro-commissioning	All buildings of a certain size must have an energy audit and test the building's system.	Energy
		Lighting upgrades and sub-metering	All commercial buildings must have lighting upgrades and sub-metering by 2025.	Energy
		Voluntary residential and school organic waste collection pilot	The local law requires the Department of Sanitation to initiate a pilot voluntary organic waste kerbside and school collection system.	Food
		Zero Waste Commercial Organics Rules	All large-scale commercial food establishments must start separating organic waste from mid-2016 onwards.	Food

Table 4.1 (continued)

Tool	Tool type	Policy title	Description	WEF sectors addressed
	Public education and skills development	Food sector job training	Grants are provided to small business owners, including agricultural and food related, to help them and their employees succeed.	Food
	Public–private partnerships	Mayor's Water Challenge to hotels	The challenge is designed to encourage hotels to voluntarily reduce water consumption with the city providing support.	Water
		Mayor's Water Challenge to restaurants	The challenge is designed to encourage a select group of restaurants to voluntarily reduce water consumption with the city providing support.	Water, energy
		Watershed Agricultural Programme	The city is working with the Watershed Agricultural Council to better manage nutrients, improve on-farm water quality, and sell more locally produced food that preserves water quality.	Water, food
		NYC Carbon Challenge	To follow the city's lead in reducing GHG emissions, the city encourages participants to reduce their emissions.	Energy

(continued)

Table 4.1 (continued)

Tool	Tool type	Policy title	Description	WEF sectors addressed
		Food Waste Challenge	The challenge encourages restaurants to reduce food waste with the city providing support including workshops, training, and technical expertise.	Food
		Zero Waste Challenge	The challenge was initiated to help businesses affected by the Zero Waste Commercial Organics Rules increase their recycling and organics separation and diversion efforts. Participants were provided with support including training materials, waste tracking reports.	Food
	Stakeholder participation	NYC School gardens initiative	The programme inspires, facilitates, and promotes the creation of a school garden in every public school in New York City.	Food
		NYCHA's Garden and Greening Programme	The programme supports food-producing gardens and urban farms managed by local community partners.	Food

(continued)

Table 4.1 (continued)

Tool	Tool type	Policy title	Description	WEF sectors addressed
	Information and awareness-raising	Monitoring water consumption	Customers can view water usage and bills and receive an alert for any leak detected.	Water
		NYC Retrofit Accelerator	Provides free technical assistance and advisory services for building owners to 'go green' through energy efficiency, water conservation, and clean energy upgrades.	Energy, water
		NYC Multifamily Solar Guide	Informative guide on the benefits of solar and what is required to complete projects.	Energy
	Resource mapping	NYC Solar Map	A free, interactive tool to help New Yorkers estimate their building's solar potential, savings, and pay-back period.	Energy
		NYC Energy & Water Performance Map	The map shows New Yorkers how their buildings compare to other similar buildings in terms of energy and water management.	Energy, water
	Public procurement	SchoolFood	The city has made the procurement of local and fresh food a priority in schools.	Food

(continued)

Table 4.1 (continued)

Tool	Tool type	Policy title	Description	WEF sectors addressed
	Demonstration projects	Innovative Demonstrations for Energy Adaptability	The city engages vendors of emerging and underutilised energy technologies to test their solutions in city buildings.	Energy
		Organic waste to energy demonstration project	Pure food waste from restaurants and other food outlets will be used to generate energy.	Food, energy
	Awards and public recognition	Annual water art and poetry contest	The contest for schoolchildren raises awareness of the city's shared water resources and the importance of water conservation.	Water

Notes

1. NYCEDC. 2017a. *January 2017 economic snapshot: Making sense of the number* [Online]. Available: https://www.nycedc.com/economic-data/january-2017-economic-snapshot.
2. NYC Department of City Planning. 2017. *Current and projected populations* [Online]. Available: http://www1.nyc.gov/site/planning/data-maps/nyc-population/current-future-populations.page.
3. NYC Environmental Protection. 2014. Water demand management plan. Available: http://www.nyc.gov/html/dep/pdf/conservation/water-demand-management-plan-single-page.pdf.
4. NYCEDC. 2015. Economic snapshot: A summary of New York City's economy April 2015. Available: https://www.nycedc.com/sites/default/files/files/economic-snapshot/Economic_Snapshot_April_2015_Energy_and_Environment.pdf.
5. NYCEDC. 2013. Economic snapshot: A summary of New York City's economy July 2013. Available: https://www.nycedc.com/sites/default/files/files/economic-snapshot/July%20Econ%20Snap%202013_0.pdf.
6. NYCEDC. 2015. Economic snapshot: A summary of New York City's economy April 2015. Available: https://www.nycedc.com/sites/default/files/files/economic-snapshot/Economic_Snapshot_April_2015_Energy_and_Environment.pdf.
7. Kennedy, C. A., Stewart, I., Facchini, A., Cersosimo, I., Mele, R., Chen, B., Uda, M., Kansal, A., Chiu, A., Kim, K.-G., Dubeux, C., Lebre La Rovere, E., Cunha, B., Pincetl, S., Keirstead, J., Barles, S., Pusaka, S., Gunawan, J., Adegbile, M., Nazariha, M., Hoque, S., Marcotullio, P. J., González Otharán, F., Genena, T., Ibrahim, N., Farooqui, R., Cervantes, G. & Sahin, A. D. 2015. Energy and material flows of megacities. *Proceedings of the National Academy of Sciences,* 112, 5985–5990.
8. NYCEDC. 2016. Five borough food flow: 2016 New York City food distribution & resiliency study results. Available: https://www.nycedc.com/system/files/files/resource/2016_food_supply-resiliency_study_results.pdf.
9. City of New York. 2016b. Food metrics report 2016. Available: http://www1.nyc.gov/assets/foodpolicy/downloads/pdf/2016-Food-Metrics-Report.pdf.
10. Matt Barron, B. G., Claire Ho, Rebecca Hudson, Dana Kaplan, Erica, Keberle, C. N., Caren Perlmutter, Zachary Suttile, Cameron Thorsteinson, & Deborah Tsien, L. W., Meghan Wilson. 2010. Understanding New York

City's food supply. *Prepared for New York City Mayor's Office of Long-Term Planning and Sustainability* [Online]. Available: http://mpaenvironment. ei.columbia.edu/files/2014/06/UnderstandingNYCsFoodSupply_May2010.pdf.

11. Food Bank for New York City. 2016. *Fast facts* [Online]. Available: http://www.foodbanknyc.org/newsroom/fast-facts.
12. New York City DEP. 2014. Water demand management plan. Available: http://www.nyc.gov/html/dep/pdf/conservation/water-demand-management-plan-spread.pdf.
13. New York City DEP. 2015. Water demand management report. Available: http://www.nyc.gov/html/dep/pdf/conservation/water_conservation_report2014.pdf.
14. New York City DEP. 2014. Water demand management plan. Available: http://www.nyc.gov/html/dep/pdf/conservation/water-demand-management-plan-spread.pdf.
15. Ibid.
16. New York City DEP. 2015. Water demand management report. Available: http://www.nyc.gov/html/dep/pdf/conservation/water_conservation_report2014.pdf.
17. Ibid.
18. Watershed Agricultural Council. 2016. Watershed agricultural program 2015 annual report and 2016 workload. Available: http://www.nycwatershed.org/watershed-agricultural-program-2015-annual-report-2016-workload/.
19. New York City Mayor's Office of Sustainability. 2017. *Solar panel tax abatement* [Online]. Available: http://www.nyc.gov/html/gbee/html/incentives/solar.shtml.
20. City of New York. 2009. Local Law 84 (LL84): Benchmarking Available: http://www.nyc.gov/html/planyc2030/downloads/pdf/ll84of2009_benchmarking.pdf.
21. City of New York. 2016c. *Green buildings and energy efficiency—LL84: Benchmarking* [Online]. Available: http://www.nyc.gov/html/gbee/html/plan/ll84.shtml.
22. City of New York. 2016d. *Green buildings and energy efficiency—LL85: NYC Energy Conservation Code (NYCECC)* [Online]. Available: http://www.nyc.gov/html/gbee/html/plan/ll85.shtml.
23. City of New York. 2016e. *Green buildings and energy efficiency—LL87: Energy Audits & Retro-commissioning* [Online]. Available: http://www.nyc.gov/html/gbee/html/plan/ll87.shtml.

24. City of New York. 2016f. *Green buildings and energy efficiency—LL88: Lighting Upgrades & Sub-metering* [Online]. Available: http://www.nyc. gov/html/gbee/html/plan/ll88.shtml.
25. City of New York. 2014. One city built to last: Transforming New York City's buildings for a low-carbon future. Available: http://www.nyc.gov/ html/builttolast/assets/downloads/pdf/OneCity.pdf.
26. City of New York. 2013b. New York City Mayor's carbon challenge progress report. Available: http://www.nyc.gov/html/gbee/downloads/ pdf/mayors_carbon_challenge_progress_report.pdf.
27. City of New York. 2016h. NYC carbon challenge handbook for multifam-ily buildings. Available: http://www.nyc.gov/html/gbee/downloads/pdf/ NYC-Carbon-Challenge-Handbook-for-Universities-and-Hospitals.pdf.
28. City of New York. 2016a. *About the NYC retrofit accelerator* [Online]. Available: https://retrofitaccelerator.cityofnewyork.us/about.
29. City of New York. 2015c. *Mayor de Blasio launches retrofit accelerator, providing key support for buildings to go green as NYC works toward 80x50* [Online]. Available: http://www1.nyc.gov/office-of-the-mayor/ news/651-15/mayor-de-blasio-launches-retrofit-accelerator-providing-key-support-buildings-go-green-as.
30. City of New York. 2014. One city built to last: Transforming New York City's buildings for a low-carbon future. Available: http://www.nyc.gov/ html/builttolast/assets/downloads/pdf/OneCity.pdf.
31. Ibid.
32. CUNY. 2017. *NYC solar partnership* [Online]. Available: http://www. cuny.edu/about/resources/sustainability/solar-america/partners.html.
33. NYCEDC. 2017b. *NYC solar partnership* [Online]. Available: https:// www.nycedc.com/program/nyc-solar-partnership.
34. City of New York. 2016g. *New York City energy & water performance map* [Online]. Available: http://benchmarking.cityofnewyork.us/.
35. City of New York. 2015b. *Mayor de Blasio announces major progress in greening city buildings* [Online]. Available: http://www1.nyc.gov/office-of-the-mayor/news/933-15/mayor-de-blasio-major-progress-greening-city-buildings.
36. City of New York. 2015a. Food metrics report 2015. Available: http:// www1.nyc.gov/assets/foodpolicy/downloads/pdf/2015-food-metrics-report.pdf.
37. City of New York. 2013a. Mayor's Food Waste Challenge to restaurants. Available: http://www.nyc.gov/html/sbs/downloads/pdf/neighborhood_ development/nddblog/FWC_overview.pdf.

38. City of New York. 2015a. Food metrics report 2015. Available: http://www1.nyc.gov/assets/foodpolicy/downloads/pdf/2015-food-metrics-report.pdf.
39. Ibid.
40. Ibid.
41. Ibid.
42. New York City Department of Sanitation. 2014. Diversion report II NYC organics collection pilot.
43. Ibid.
44. New York City Department of Sanitation. 2015. Notice of public hearing and opportunity to comment on proposed rules. Available: https://www1.nyc.gov/assets/dsny/docs/zerowaste_DSNY-Proposed-Commercial-Organics-Rule-With-Certifications-Preliminary-Certifications-8-19-15_0915.pdf.
45. City of New York. 2016i. *NYC Mayor's Zero Waste Challenge* [Online]. Available: http://www1.nyc.gov/site/sustainability/initiatives/zero-waste-challenge.page.

References

City of New York. 2009. Local Law 84 (LL84): Benchmarking. http://www.nyc.gov/html/planyc2030/downloads/pdf/ll84of2009_benchmarking.pdf

———. 2013a. Mayor's Food waste challenge to restaurants. http://www.nyc.gov/html/sbs/downloads/pdf/neighborhood_development/nddblog/FWC_overview.pdf

———. 2013b. New York City Mayor's carbon challenge progress report. http://www.nyc.gov/html/gbee/downloads/pdf/mayors_carbon_challenge_progress_report.pdf

———. 2014. One city built to last: Transforming New York City's buildings for a low-carbon future. http://www.nyc.gov/html/builttolast/assets/downloads/pdf/OneCity.pdf

———. 2015a. Food metrics report 2015. http://www1.nyc.gov/assets/foodpolicy/downloads/pdf/2015-food-metrics-report.pdf

———. 2015b. *Mayor de Blasio announces major progress in greening city buildings* [Online]. http://www1.nyc.gov/office-of-the-mayor/news/933-15/mayor-de-blasio-major-progress-greening-city-buildings

———. 2015c. *Mayor de Blasio launches retrofit accelerator, providing key support for buildings to go green as NYC works toward 80x50* [Online].

http://www1.nyc.gov/office-of-the-mayor/news/651-15/mayor-de-blasio-launches-retrofit-accelerator-providing-key-support-buildings-go-green-as

———. 2016a. *About the NYC retrofit accelerator* [Online]. https://retrofitaccelerator.cityofnewyork.us/about

———. 2016b. Food metrics report 2016. http://www1.nyc.gov/assets/food-policy/downloads/pdf/2016-Food-Metrics-Report.pdf

———. 2016c. *Green buildings and energy efficiency—LL84: Benchmarking* [Online]. http://www.nyc.gov/html/gbee/html/plan/ll84.shtml

———. 2016d. *Green buildings and energy efficiency—LL85: NYC Energy Conservation Code (NYCECC)* [Online]. http://www.nyc.gov/html/gbee/html/plan/ll85.shtml

———. 2016e. *Green buildings and energy efficiency—LL87: Energy Audits & Retro-commissioning* [Online]. http://www.nyc.gov/html/gbee/html/plan/ll87.shtml

———. 2016f. *Green buildings and energy efficiency—LL88: Lighting Upgrades & Sub-metering* [Online]. http://www.nyc.gov/html/gbee/html/plan/ll88.shtml

———. 2016g. *New York City energy & water performance map* [Online]. http://benchmarking.cityofnewyork.us/

———. 2016h. NYC carbon challenge handbook for multifamily buildings. http://www.nyc.gov/html/gbee/downloads/pdf/NYC-Carbon-Challenge-Handbook-for-Universities-and-Hospitals.pdf

———. 2016i. *NYC Mayor's Zero Waste Challenge* [Online]. http://www1.nyc.gov/site/sustainability/initiatives/zero-waste-challenge.page

CUNY. 2017. *NYC solar partnership* [Online]. http://www.cuny.edu/about/resources/sustainability/solar-america/partners.html

Food Bank for New York City. 2016. *Fast facts* [Online]. http://www.foodbanknyc.org/newsroom/fast-facts

Kennedy, C.A., I. Stewart, A. Facchini, I. Cersosimo, R. Mele, B. Chen, M. Uda, et al. 2015. Energy and material flows of megacities. *Proceedings of the National Academy of Sciences* 112: 5985–5990.

Matt Barron, B.G., Rebecca Hudson Claire Ho, Erica Keberle Dana Kaplan, Caren Perlmutter Cullen Naumoff, Cameron Thorsteinson, Zachary Suttile, L.W. Deborah Tsien, and Meghan Wilson. 2010. Understanding New York City's food supply. *Prepared for New York City Mayor's Office of Long-Term Planning and Sustainability* [Online]. http://mpaenvironment.ei.columbia.edu/files/2014/06/UnderstandingNYCsFoodSupply_May2010.pdf

New York City DEP. 2014. Water demand management plan. http://www.nyc.gov/html/dep/pdf/conservation/water-demand-management-plan-spread.pdf

————. 2015. Water demand management report. http://www.nyc.gov/html/dep/pdf/conservation/water_conservation_report2014.pdf

New York City Department of Sanitation. 2014. Diversion report II NYC organics collection pilot.

————. 2015. Notice of public hearing and opportunity to comment on proposed rules. https://www1.nyc.gov/assets/dsny/docs/zerowaste_DSNY-Proposed-Commercial-Organics-Rule-With-Certifications-Preliminary-Certifications-8-19-15_0915.pdf

New York City Mayor's Office of Sustainability. 2017. *Solar panel tax abatement* [Online]. http://www.nyc.gov/html/gbee/html/incentives/solar.shtml

NYC Department of City Planning. 2017. *Current and projected populations* [Online]. http://www1.nyc.gov/site/planning/data-maps/nyc-population/current-future-populations.page

NYC Environmental Protection. 2014. Water demand management plan. http://www.nyc.gov/html/dep/pdf/conservation/water-demand-management-plan-single-page.pdf

NYCEDC. 2013, July. Economic snapshot: A summary of New York City's economy. https://www.nycedc.com/sites/default/files/files/economic-snapshot/July%20Econ%20Snap%202013_0.pdf

————. 2015, April. Economic snapshot: A summary of New York City's economy. https://www.nycedc.com/sites/default/files/files/economic-snapshot/Economic_Snapshot_April_2015_Energy_and_Environment.pdf

————. 2016. Five borough food flow: 2016 New York City food distribution & resiliency study results. https://www.nycedc.com/system/files/files/resource/2016_food_supply-resiliency_study_results.pdf

————. 2017a. *January 2017 economic snapshot: Making sense of the number* [Online]. https://www.nycedc.com/economic-data/january-2017-economic-snapshot

————. 2017b. *NYC solar partnership* [Online]. https://www.nycedc.com/program/nyc-solar-partnership

Watershed Agricultural Council. 2016. Watershed agricultural program 2015 annual report and 2016 workload. http://www.nycwatershed.org/watershed-agricultural-program-2015-annual-report-2016-workload/

5

The Green Economy and the Water-Energy-Food Nexus in Singapore

Introduction

Since independence in 1965, Singapore's economy has evolved from exporting labour-intensive products to developing high value-added products including electronics, chemicals, and biomedical. Over the period 2000–2010, the GDP nearly doubled from S$163 billion to S$304 billion.[1] However, currently Singapore's economic growth is between 1 and 2 percent.[2] The city-state's population grew by 1.3 percent in 2015 to reach 5.61 million in 2016.[3]

5.1 Water-Energy-Food Nexus Pressures

Singapore is experiencing a variety of water, energy, and food nexus pressures that are detrimental to the development of a green economy as described below through a variety of examples.

© The Author(s) 2018
R.C. Brears, *The Green Economy and the Water-Energy-Food Nexus*,
DOI 10.1057/978-1-137-58365-9_5

Water

It is projected that Singapore will be one of the most water-stressed countries in the world. By 2040, Singapore could be one of eight countries in the world most vulnerable to disruptions in water supply.[4] Currently, water demand in Singapore is around 430 million gallons per day, with homes consuming 45 percent and the non-domestic sector consuming the rest. However, by 2060 total water demand could almost double with the non-domestic sector consuming about 70 percent of the water. By then, recycled water and desalination will meet up to 85 percent of Singapore's future water demand.[5] If Singapore relies on current desalination technology, energy use will be four times greater than now.[6]

Energy

Singapore's total electricity generation increased by 2 percent from 2014 to 2015, with natural gas constituting 95 percent of the fuel mix. Meanwhile total electricity consumption rose by 2.4 percent in 2015 due to industrial-related consumption, which comprises 42 percent of total electricity consumption in 2015, followed by the commerce and services-related sector and households consuming 37 percent and 15 percent of consumption respectively.[7]

Food

When Singapore gained independence in 1965, it was partially self-sufficient in food. Approximately 20,000 farms occupied more than 25 percent of the land, supplying up to 60 percent of Singapore's vegetable needs. The land was occupied by small-scale, family-owned farms for food and plantations. However, over time the government steered the economy towards industrialisation to create jobs and growth with small farms reallocated to make way for planned new towns with high-density housing and industry. The result is that today more than 90 percent of Singapore's food is imported while farms occupy less than 1 percent of Singapore's land. To ensure Singapore's food is not vulnerable to disruptions or impacts from any new single sources, the government is working with industry and markets to diversify food sources including increasing local production.[8]

5.2   General Fiscal Tools to Reduce Water-Energy-Food Nexus Pressures

Singapore has implemented a variety of fiscal tools that create interdependencies and synergies between the nexus systems while reducing trade-offs between the systems in the development of a green economy.

HDB Greenprint Fund

The Housing and Development Board's (HDB) Greenprint Fund will fund and support test-bedding of smart and sustainable solutions (new and original, encompass workable or implementable products, systems, processes, or work methodologies) (Table 5.1) within the pilot neighbourhood of Yuhua, subject to a cap of S$100,000 per application. HDB will disburse the fund through reimbursement of the total qualifying cost incurred in the development or test-bedding process.[9]

5.3   General Non-fiscal Tools to Reduce Water-Energy-Food Nexus Pressures

Singapore has implemented a variety of non-fiscal tools that create interdependencies and synergies between the nexus systems while reducing trade-offs between the systems in the development of a green economy.

Table 5.1 HDB Greenprint Fund smart sustainable solutions

Solutions	Description
Waste	Reduce waste and increase recycling
Building/infrastructure	Reduce maintenance and material wastage
Water	Save water
Environment	Improve environment and quality of living
Air	Improve air quality in the environment
Energy	Save energy
Community	Improve quality of life and foster community bonding

Housing and Development Board. 2017. *HDB Greenprint fund* [Online]. Available: http://www.hdb.gov.sg/cs/infoweb/about-us/our-role/smart-and-sustainable-living/hdb-greenprint/funding

Green Mark for Buildings

In 2005, the Building and Construction Authority (BCA) launched the BCA Green Mark scheme, a three-tiered rating system (Certified Gold, Gold, and Platinum) to evaluate a building's environmental impact and recognise its sustainability performance. The goal of the scheme is that by 2030, 80 percent of all buildings achieve the standards of Green Mark, up from around 25 percent today. Certified Green Mark buildings are required to be re-assessed every three years to maintain the Green Mark status. For new buildings, the Green Mark scheme encourages developers and design teams to design and construct green, sustainable buildings that promote energy and water savings, healthier indoor environments as well as the adoption of more extensive greenery for their projects. For existing buildings, the scheme encourages building owners and operators to meet sustainable operational goals and reduce adverse impacts of their buildings on the environment and occupant health over the entire building lifecycle.[10]

Expanding the Green Mark Scheme Beyond Buildings

In 2010, the Green Mark scheme was revised to include a range of occupant-centric Green Mark certifications to promote Green Mark in office interiors, restaurants, supermarkets, retail outlets, and data centres (summarised in Table 5.2) as well as parks, supporting infrastructure, districts, and rapid transit systems.

Green Mark Pearl Award

In 2014, the BCA launched the Green Mark Pearl award to recognise the strong commitment of developers, building owners, landlords, and tenants of the same project/building who work in tandem to achieve greater environmental sustainability in their project/building. The award is given to developers, building owners, and landlords who have a significant number of tenants who are Green Mark certified under the Green Mark

Table 5.2 Occupant-centric Green Mark certification programmes

Green Mark certification programme	Aim of certification
BCA Green Mark for office interiors	Promotes the use of environmentally friendly features and sustainable practices by office tenants, for example, setting the office temperature no lower than 24 °C and monitoring and maintaining good air quality
BCA Green Mark for restaurants	Promotes corporate responsibility as well as recognises environmentally friendly and sustainable practices in restaurants, for example, using energy-efficient kitchen equipment and environmentally friendly materials for food containers and carry-out bags
BCA Green Mark for supermarket	Promotes environmentally friendly as well as sustainable practices and features in supermarket operations, for example, operators reducing, reusing, and recycling items including cardboard and implementation of 'green checkout lanes' for consumers with their own recycling bags
BCA Green Mark for retail	Recognises the efforts of individual retail tenants for their sustainability efforts, for example, the installation of energy-efficient lighting
BCA Green Mark for data centre	Assesses data centres on their energy efficiency, sustainable construction and management, indoor environmental quality, and other 'green' features
BCA Green Mark for healthcare facilities	Encourages the adoption of resource- and energy-efficient design, technologies, and features among healthcare facilities

Building and Construction Authority. 2013b. Singapore leading the way for green buildings in the tropics. Available: https://www.bca.gov.sg/greenmark/others/sg_green_buildings_tropics.pdf

occupant-centric schemes within a base building that is Green Mark Gold or higher. There are two tiers of the award: Green Mark Pearl Prestige award and Green Mark Pearl award that are given out to three types of buildings: commercial offices, retail malls, and business park developments. Each year there is only one winner of the Green Mark Pearl Prestige award and five winners of the Green Mark Pearl award. Overall, the award aims to:

- Emphasise the importance of total building performance of the building/project.

- Recognise developers, building owners, and landlords who have made an effort and taken active steps to work with tenants to shape their behaviour and operational practices.
- Encourage developers and building owners to ensure their green building is operating and performing as designed, both on the outside, that is, common areas and building services within the developer/building owner/landlord's control, and inside, that is, tenanted spaces.
- Recognise developers, building owners, and landlords who demonstrate thought leadership.[11]

Expanding the Green Mark Criteria

In 2015, the criteria for Green Mark were restructured into four main categories: climatic response, building energy performance, resource stewardship, and smart and healthy building. In addition, a bonus section was added for advanced green building efforts to spur efforts beyond requirements (Table 5.3). The overall sustainability outcomes of Green Mark 2015 are:

- *Climate responsiveness*: Building designs are optimised for site-specific and climatic conditions.
- *Resource stewardship*: Building designs optimise/minimise consumption of the world's resources throughout the buildings' lifecycle.
- *Enhanced health and well-being*: Building systems and operations take care of human health and people's well-being.
- *Conservation and ecological systems*: Building designs incorporate measures that conserve ecosystems and preserve abiotic cycles.

Training Green Specialists

The BCA Academy has developed a training framework to build the industry's capability in the whole design chain of green building design, construction, maintenance, and management. In particular, the BCA Academy has developed executive, academic, and specialist training programmes in collaboration with local and international educational providers to train green building professionals, specialists, and experts in the design and management of sustainable buildings.

Table 5.3 Green Mark 2015 criteria and indicators

Criteria	Indicators	Description
Climatic response	Leadership	• Climatic and contextually responsive design brief • Integrative design process • Environmental credentials of project team • Building information modelling • Green users
	Urban harmony	• Sustainable urbanism • Integrated landscape and waterscape
	Tropicality	• Tropical façade performance • Internal organisation • Ventilation performance
Building energy performance	Energy efficiency	• Air conditioning system efficiency • Lighting efficiency
	Energy effectiveness	• Building energy • Carpark energy • Receptacle energy
	Renewable energy	• Feasibility study • Solar ready roof • Replacement energy
Resource stewardship	Water	• Water efficiency system • Water monitoring and leak detection system • Alternative sources of water
	Materials	• Sustainable construction • Embodied energy and life cycle • Sustainable products
	Waste	• Environmental construction management programmes • Operational waste management • Provision of recycling system for horticulture or food waste
Smart and healthy building	Smart operations	• Energy monitoring • Demand control • Integration and analytics • System handover and documentation
	Spatial quality	• Lighting • Acoustics • Well-being
	Indoor air quality	• Occupant comfort • Outdoor air • Contaminants
Advanced green efforts	Cost-effective design Complementary certifications Social benefits	

Building and Construction Authority. 2015. Build green

Green School Roadmap

The BCA has developed a Green School Roadmap that gives younger generations first-hand experience in sustainability issues. The roadmap aims to green and facilitate schools in achieving overall net zero energy. This will be achieved by helping schools conduct energy audits of their premises through the Greenovate Challenge programme; attaining a BCA Green Mark for existing schools award; and installing PV panels in schools to achieve net zero energy.[12]

Achieving Minimum Environmental Sustainability Standards for Existing Buildings

Singapore's Building Control Act requires that from 2 January 2014 building owners achieve the minimum environmental sustainability standard (Green Mark Standard) for existing buildings when installing or retrofitting a cooling system as well as carry out periodic energy audits on the building's cooling system and comply with design system efficiency standards. This is in addition to all building owners having to submit from 1 July 2013 building information and energy consumption data annually to BCA's Building Energy Submission System. This allows the BCA to monitor building energy consumption and encourage building occupants and users to be proactive in improving building energy efficiency and reducing energy consumption.[13]

Carbon Footprint Assessment Toolkit

The BCA has been working with non-governmental organisations and academic researchers to develop a carbon footprint assessment toolkit for buildings. In addition, industry training is being developed to meet new BCA Green Mark standards requiring: quantification of the carbon footprint of key components of building projects, ability to conduct sensitivity analysis, and carbon footprint reporting through established reporting standards.[14]

Working with Industry to Enhance Indoor Environmental Quality

The BCA is working with industry to review various building codes involving ventilation, lighting as well as health and safety to ensure good indoor environmental quality. In addition, the Green Mark certification process will provide provisions to ensure facilities use only products that do not release harmful gases as well as have in place measures to conduct regular audits of the buildings' indoor air quality.[15]

5.4   Water: Fiscal Tools

Singapore has implemented a variety of water-related fiscal tools that create interdependencies and synergies between the nexus systems while reducing trade-offs between the systems in the development of a green economy.

Potable Water Price Revisions

Water prices will be revised in Singapore from 1 July 2017 with tariffs going up in two steps: in July 2017 and in July 2018. A range of investments in water infrastructure, along with rising operational costs, has made the increase in water prices necessary. This will ensure Singapore can meet future demand, increase water security, and continue to deliver a high quality and reliable supply of water. The key revisions (Table 5.4) are:

- A 30 percent increase in water price, phased over two years, starting from 1 July 2017
- Restructuring of the Sanitary Appliance Fee and the Waterborne Fee into a single, volume-based fee

NEWater Price Revisions

The price of NEWater—high-grade reclaimed water produced from treated water which is further purified using advanced membrane

Table 5.4 Singapore's potable water price revisions

		Current Water prices ($/m³)		From 1 July 2017 Water prices ($/m³)		From 1 July 2018 Water prices ($/m³)	
		0–40 m³	>40 m³	0–40 m³	>40 m³	0–40 m³	>40 m³
Potable water	Tariff	$1.17	$1.40	$1.19	$1.46	$1.21	$1.52
	Water conservation tax	$0.35 (30% of $1.17)	$0.63 (45% of $1.40)	$0.42 (35% of $1.19)	$0.73 (50% of $1.46)	$0.61 (50% of $1.21)	$0.99 (65% of $1.52)
Used water	Waterborne fee	$0.28	$0.28	$0.78	$1.02	$0.92	$1.18
	Sanitary appliance fee	$2.80 per fitting	$2.80 per fitting	Combined into waterborne fee		Combined into waterborne fee	
Total price		$2.10	$2.61	$2.39	$3.21	$2.74	$3.69

PUB. 2017g. *Water prices* [Online]. Available: https://www.pub.gov.sg/PublishingImages/T1A%20dom.jpg

Table 5.5 NEWater price revisions

		Current Water prices ($m³)	From 1 July 2017 Water prices ($m³)	From 1 July 2018 Water prices ($m³)
NEWater	Tariff	$1.22	$1.28	$1.28
	Water conservation tax (% of NEWater tariff)	–	$0.13 (10% of $1.28)	$0.13 (10% of $1.28)
Used water	Waterborne fee	$0.56	$0.78	$0.92
Total price		$1.78	$2.19	$2.33

PUB. 2017c. *Industrial water prices* [Online]. Available: https://www.pub.gov.sg/PublishingImages/T5A%20NEWater.jpg

technologies and ultraviolet disinfection—will be revised over the next two years (Table 5.5), with key revisions being:

• From July 2017, an increase in the NEWater tariff and a 10 percent Water Conservation Tax imposed on NEWater
• The increase in the Waterborne Fee will be phased in over two years, from July 2017 and in July 2018

Water Efficiency Fund

The Public Utilities Board's (PUB) Water Efficiency Fund (WEF) encourages large water-using organisations to seek efficient ways to manage their water demands. Organisations that need to comply with the Water Efficiency Management Plan (WEMP) can tap into the WEF to implement water-saving measures if they can meet the fund's criteria. Commercial water users who develop their WEMP voluntarily can apply to the fund to finance the installation of water meters.[16]

WEF co-funds projects that yield at least a 10 percent reduction in water consumption within their organisations. Companies organising community-wide campaigns are also eligible for funding. Projects eligible for funding are summarised in Table 5.6.

5.5   Water: Non-fiscal Tools

Singapore has implemented a variety of water-related non-fiscal tools that create interdependencies and synergies between the nexus systems while reducing trade-offs between the systems in the development of a green economy.

Mandatory Water Efficiency Labelling Scheme

In 2009, Singapore's PUB introduced the Mandatory Water Efficiency Labelling Scheme (MWELS) to help consumers make informed purchasing decisions and encourage suppliers to introduce more water-efficient products. Under this scheme suppliers are required to label the water efficiency of their water fittings and appliances on all displays, packaging and advertisements. Since 2009, MWELS has covered taps, urinal flush valves, waterless urinals and dual-flush low capacity flushing cisterns. In 2011 MWELS was extended to cover washing machines. In addition, minimum water efficiency standards were introduced in 2009 requiring all new developments and existing premises undergoing renovation to install water fittings with at least a

Table 5.6 Water Efficiency Fund projects

Project	Funding
SS 577:2012 Water efficiency management systems certification	PUB co-funding 50 percent of the total costs of certification capped at S$10,000 following a feasibility study project, which itself will be 50 percent funded, subject to a cap of S$50,000
Water audit project	PUB will co-fund 90 percent of the water audit cost, subject to a cap of S$15,000
Recycling efforts/use of alternative sources of water	PUB will support customers at S$0.60 for every cubic metre of potable water and S$0.30 for every cubic metre of new water (reclaimed water) and S$0.10 for every cubic metre of industrial water (industrial water recycled on-site) saved over the economic life of the facilities or seven years (whichever is earlier), up to S$1 million
Community-wide water conservation campaigns and programmes	PUB will co-fund 50 percent of the cost of community-wide water conservation programmes, subject to a cap that ranges from S$200 to S$5000 depending on the type of programme

PUB. 2016a. Application for Water Efficiency Fund (WEF). Available: https://www.pub.gov.sg/Documents/WEF_ApplicationForm_2016.pdf; PUB. 2017b. *Efficiency measures* [Online]. Available: https://www.pub.gov.sg/savewater/atwork/efficiencymeasures

'1-tick' water efficiency rating. Since October 2015, only 2- and 3-tick WELS rated washing machines are allowed to be sold in Singapore.[17, 18]

Water Efficiency Management Plan

Since 2015, PUB requires all large water users with water consumption of at least 60,000 cubic metres in the previous year to:

- Submit to PUB by 31 March a notification of the different sites that meet this consumption threshold
- Install private water meters at various water usage areas within their premises by 30 June to track and monitor water usage
- Submit an annual WEMP by 30 June for at least three consecutive years

The Water Efficiency Management Plan

The Water Efficiency Management Plan allows customers to:

- Methodologically break down water usage in the premises and includes a water balance chart, which compares total water supplied to the site, the actual water consumed within all the water end uses on-site and the total water leaving the site. This helps identify areas of significant water use and problem areas including leaks.
- Identify areas to further reduce consumption and raise efficiency.
- Establish an action plan which identifies measures in water savings, priorities, and implementation timelines.[19,20]

Water Efficient Building Certification

PUB has implemented a Water Efficient Building (WEB) certification that focuses on water usage in individual buildings. The WEB certification encourages non-domestic buildings, for example those run by businesses, industries, and schools, to put in place water-efficient measures. A building can be awarded WEB (Basic) if it installs water-efficient fittings or adopts water-efficient flow rates or flush volumes that save around 5 percent of its monthly water consumption. Buildings that are exemplary performers in water efficiency and who the adopt water efficiency management system—SS 577:2012 Water Efficiency Management Systems—can achieve WEB (Silver) or WEB (Gold) certification (summarised in Table 5.7).

Table 5.7 Water Efficient Building certification criteria

Water Efficient Building (Basic)	Water Efficient Building (Silver)	Water Efficient Building (Gold)
Installed water-efficient fittings or adopted water-efficient flow rates/flush volumes	Attained Water Efficient Building certification (Basic)	
	Installed private water meters to monitor main usage areas	
	Completed and submitted WEMP to PUB (non-domestic users with water consumption less than 60,000 cubic metres/year do not have to submit their WEMP to PUB; however, if they are applying for WEB Certification, then they have to complete and submit WEMP to PUB as part of the certification's criteria)	
	Installed fault-reporting system to report leaks, faulty fittings, etc.	
	Water consumption monitoring: At least monthly	Water consumption monitoring: At least weekly
	Water efficiency index or recycling rate	
	Top 25 percent on water efficiency performance	Top 10 percent on water efficiency performance
		Certified to SS 577:2012 Water efficiency management systems

PUB. 2017f. *Water Efficient Building certification programme* [Online]. Available: https://www.pub.gov.sg/savewater/atwork/certificationprogramme

10 Percent Challenge for Non-domestic Users

PUB has established the 10 percent challenge for non-domestic customers to reduce their monthly water consumption by 10 percent. Through a web portal and the WEB programme customers can access cost-effective solutions.

Water Efficient Building Design Guide Book

PUB has also developed the WEB design guide book that provides guidelines and recommendations for adoption by the industry in designing for water efficiency in building developments.

Water Conservation Education

PUB has worked with the Ministry of Education to incorporate water conservation topics in the social studies syllabus for Primary 3 students. In addition, PUB has developed a 'Time to Save' water conservation programme for Primary 3 students in which all participants use a timer and activity booklet to track their shower timings for a week. During the week the students will also become junior water advocates encouraging family members to take shorter showers as well as share water conservation tips with their neighbours.[21]

Water Efficient Homes Programme

The Water Efficient Homes programme helps residents save water at home and reduce their water bills. The programme encourages residents to install water-saving devices and practice good water conservation habits. As part of the programme, PUB officers visit households to install water-saving devices free of charge.

5.6   Energy: Fiscal Tools

Singapore has implemented a variety of energy-related fiscal tools that create interdependencies and synergies between the nexus systems while reducing trade-offs between the systems in the development of a green economy.

Green Mark Incentive Scheme for Existing Buildings and Premises

In 2014, the BCA launched the S$50 million Green Mark Incentive Scheme for existing buildings and premises to encourage and incentivise building owners, occupants, and tenants to undertake and adopt energy-efficient improvements and measures within their buildings and premises. The incentive is available to existing small and medium enterprises (SMEs) tenants and building owners or building owners with at least 30 percent of their tenants classified as SMEs. The SMEs will be required to meet minimum of 30 percent local shareholding, and annual sales turnover of no more than S$100 million or employment size of not more than 200 employees. The scheme will co-fund up to 50 percent of the retrofitting cost for energy improvements or up to S$3 million for building owners and up to S$20,000 for occupants and tenants. In addition, the scheme will also help build up the industry's capability of undertaking retrofits in medium-sized buildings and premises.[22]

Green Mark Incentive Scheme—Design Prototype

The Green Mark Incentive Scheme—Design Prototype aims to encourage developers and building owners to place a greater emphasis on energy efficiency in the design stage. The scheme, running from 2014 to 2018, provides funding support for the engagement of Environmentally Sustainable Design consultants to conduct collaborative design workshops and assist in simulation studies early in the project to achieve optimal design for green buildings, with funding support capped at 70 percent of the costs of engaging the consultants or S$600,000, whichever is lower. To be eligible for the incentive the development should be in its preliminary concept design stage and target achieving beyond Green Mark Platinum, demonstrating energy savings of at least 40 percent better than the current base code.[23]

Energy Efficiency Improvement Assistance Scheme

The Energy Efficiency Improvement Assistance Scheme is a co-funding scheme managed by the National Environmental Agency to encourage companies to carry out energy appraisals and identify potential areas for energy efficiency improvements in their facilities. The scheme funds up to 50 percent of the costs of engaging an expert consultant or Energy Services Company to conduct energy appraisals and recommend specific measures that can be implemented to improve energy efficiency. Over a five-year period, the maximum amount of funding to any single facility or building is capped at S$200,000.[24]

Design for Efficiency Scheme

The Design for Efficiency Scheme encourages investors in new facilities and facility expansion projects to incorporate energy and resource efficiency practices early in the design stage of a facility before investment choices are locked in. The scheme involves a two- to three-day design workshop where internal experts (home team) and external experts (visiting team) are brought together to identify technical opportunities in energy and resource efficiency (e.g. water, gases), waste reduction, and reuse as well as the cost-effective supply of energy and water. Investors can receive funding of 50 percent of the design workshop fee or S$600,000, whichever is lower.[25]

Productivity Grant (Energy Efficiency)

The Grant for Energy Efficient Technologies encourages owners and operators of new and existing industrial facilities to invest in energy-efficient equipment or technologies. The grant can be used to cover up to 20 percent of the qualifying costs including manpower, equipment, and materials and professional services, capped at S$4 million per project. To be eligible for the grant the applicant must have implemented an energy management system that is comprised as follows:

- Energy policy stating the organisation's commitment to achieving energy performance improvements
- Energy objectives and energy targets along with time frames for achieving the objectives and targets

- Energy efficiency improvement plan for achieving the objectives and targets
- Roles and responsibilities that nominate an energy manager
- Monitoring, measuring, and analysis to determine significant energy uses in the organisation, relevant variables related to energy use, energy performance indicators, effectiveness of energy efficiency improvement plans in achieving objective and targets, and actual versus expected energy consumption[26]

Electronic Road Pricing

Singapore's Land Transport Authority (LTA) aims for a 'car-lite Singapore' by encouraging people to travel by sustainable modes, that is, public transport, walking, or cycling, to meet increasing travel demand and reduce emissions from the transport sector. To discourage congestion of roads the LTA maintains an Electronic Road Pricing (ERP) tool. The tool is based on a pay-as-you-use principle where motorists are charged when they use congested stretches of the road. The system is designed so motorists can choose to pay the ERP and enjoy a smooth rise, switch travelling times or routes, or use public transport. ERP rates are based on local traffic conditions and differ by locations and time to ensure roads remain within the optimal speed range, that is, 45–65 km/h for expressways and 20–30 km/h for arterial roads: considered the speed ranges that allow for optimal throughput on Singapore's roads. For instance, when speeds fall consistently below 45 km/h on an expressway the ERP rate will be increased and when speeds are consistently above 65 km/h on an expressway the ERP rate will be reduced.[27]

5.7   Energy: Non-fiscal Tools

Singapore has implemented a variety of energy-related non-fiscal tools that create interdependencies and synergies between the nexus systems while reducing trade-offs between the systems in the development of a green economy.

Green Building Innovation Cluster's Energy Efficient Demonstrations Scheme

Singapore has established the Green Building Innovation Cluster (GBIC) to increase the deployment of energy-efficient building solutions. With a fund of S$52 million GBIC aims to develop large-scale high-impact demonstration projects for promising technologies and solutions. GBIC will also facilitate the widespread adoption of energy-efficient solutions and practices developed. The GBIC Building Energy Efficient Demonstrations Scheme (GBIC-Demo) serves to demonstrate novel energy-efficient technologies that have not been widely implemented locally in actual building operations. GBIC-Demo aims to link building owners with technology providers in order to establish platforms where industry can test and showcase these technologies to generate local performance data for verification, from which it is aimed that successful technologies are replicated in other buildings and eventually commercialised.[28]

Energy Efficiency National Partnership

In 2010, the National Environmental Agency launched an industry-focused Energy Efficiency National Partnership (EENP) programme. The EENP is a voluntary partnership programme for companies that wish to be more energy efficient, which in turn enhances their long-term competitiveness while reducing their carbon footprint. The EENP supports companies in their energy efficiency efforts through energy-efficiency-related resources, learning network activities, and recognition. The EENP is targeted at companies that are interested in improving their energy efficiency and implementing energy management practices and companies that are large energy consumers, particularly ones that consume 54 TJ of energy or more per annum. The EENP programme comprises three components:

1. *Energy management system*: Partner companies are encouraged to adopt an energy management system to measure and manage energy consumption as well as identify energy efficiency improvements.

2. *EENP learning networks*: The learning network provides a forum for companies to learn about energy efficiency ideas, technologies, practices, standards, and case studies. Forums include conferences and technical workshops aimed at senior and middle management and technical staff.
3. *EENP national recognition scheme*: The EENP awards recognise efforts and achievements of corporates and corporate teams that have made significant energy savings.[29]

Mandatory Energy Management Requirements

Energy-intensive companies that consume 54 TJ of energy or more each year are required to appoint an energy manager, monitor and report energy use and GHG emissions, and submit energy efficiency improvement plans.

5.8   Food: Fiscal Tools

Singapore has implemented a variety of food-related fiscal tools that create interdependencies and synergies between the nexus systems while reducing trade-offs between the systems in the development of a green economy.

Agricultural Productivity Fund

Local farms provide a buffer in times of sudden food supply disruptions. Since 2009 Agri-Food and Veterinary Authority's (AVA) Food Fund has been co-funding farms in upgrading their production capability and supporting farms in research and development in food farming technology. Nonetheless, land is scarce in Singapore and farms need to achieve higher productivity and use farmland more efficiently. As such, AVA has initiated the new Agricultural Productivity Fund (APF) to replace the existing Food Fund. The APF consists of three components (summarised in Table 5.8):

Table 5.8 Agricultural Productivity Fund for food farms

Funding component	Funding level	Funding cap
Basic capability upgrading	50%	S$50,000
Productivity enhancement	70%	S$700,000
Research and development	70–90%	S$250,000 for proof-of-concept S$500,000 for proof-of-value S$1 million for test-bedding

Agri-Food and Veterinary Authority of Singapore. 2016. Agriculture productivity fund and agriculture policies for local farms. Available: http://app.mnd.gov.sg/Portals/0/Annex%20A_APF_AVA.pdf

1. *Basic capability upgrading*: Funding support for equipment that would increase productivity
2. *Productivity enhancement*: Funding for quantum leaps in productivity and production capability through the adoption of automated systems or other hi-tech systems
3. *Research and development*: Funding of technology for intensive farming systems, consultancy services in land intensification, and technological pilot trials to explore innovative and commercially viable solutions to maximise farm productivity

5.9   Food: Non-fiscal Tools

Singapore has implemented a variety of food-related non-fiscal tools that create interdependencies and synergies between the nexus systems while reducing trade-offs between the systems in the development of a green economy.

Intensifying Land Use

To intensify land use, local farms will have to adhere to a series of new policies:

- *Use 90 percent of farmland for production*: To ensure maximum use of farmland for production purposes and to maintain high levels of local

Table 5.9 Minimum production levels for food farms

Farm type	Minimum production levels
Vegetables	
Leafy vegetables	130 tonnes/hectare/year
Bean sprout	1000 tonnes/hectare/year
Food fish	
Land-based	25 metric tonnes/hectare/year
	Hatcheries:
	Fish fry up to 5 cm: 0.75 million/hectare/year
	Fish fry 5–10 cm: 0.5 million/hectare/year
Eggs	
Hen eggs	9,000,000 pieces/hectare/year
Quail eggs	7,000,000 pieces/hectare/year

Energy Efficient Singapore. 2017a. *Energy efficiency national partnership (EENP)* [Online]. Available: http://www.e2singapore.gov.sg/Programmes/Energy_ Efficiency_National_Partnership.aspx

production all farms must use at least 90 percent of their farm area for agricultural production and uses related to production. Only 10 percent of the land may be used for ancillary purposes including dwelling houses, workers quarters, offices, and visitor centres. This policy will be applicable to all land-based farms licensed by AVA for lease extensions and new leasers.

- *Minimum production levels*: To ensure farms use land efficiently there will be minimum production levels set for all land-based farms licensed by AVA (Table 5.9)
- *Land tenures*: AVA will tender new farmlands with ten-year leases. Farms may renew or extend their farm leases in tenures of ten years with the option to extend for another ten on condition they are able to meet the minimum production targets.[30]

Food Innovation and Resource Centre

The Food Innovation and Resource Centre (FIRC) was launched in 2007 as a joint initiative between Singapore Polytechnic and SPRING Singapore, which is an agency under the Ministry of Trade and Industry responsible for helping Singapore enterprises grow and building trust in Singapore products and services. FIRC's goal is to provide food enterprises with technical expertise in new product and process development. One of the main programme areas of FIRC is Process Engineering Innovation, which involves the process engineering team assisting food companies to improve and optimise productivity through the adoption of technology, process re-engineering, automation, and design customisation, with one of the main benefits being reduction in resource intensive processes and a reduction in waste.[31]

5.10   Case Study Summary

Singapore has implemented a variety of initiatives to create interdependencies and synergies between the nexus systems while reducing trade-offs between the systems in the development of a green economy.

General

- Singapore's HDB provides funding and support to companies wishing to test smart and sustainable solutions in a pilot neighbourhood, with funding available for solutions that save water as well as energy and reduce all types of waste.
- To encourage buildings to reduce their environmental impacts, the city has implemented the Green Mark Scheme to evaluate buildings and recognise outstanding sustainability performance. For new buildings, the scheme encourages the incorporation of sustainability features in the design phase and for existing buildings the scheme encourages owners and operators to meet sustainability goals and reduce adverse environmental impacts from their buildings. Since the scheme's inception certification has been expanded to include a wide variety of occupant-centric facilities including offices, retail outlets, parks, and infrastructure.
- To increase the number of green buildings in Singapore, a training programme has also been devised to train green building professionals, specialists, and experts.

Water

- To recognise growing water scarcity and increasing operational costs, Singapore has revised its water prices. The prices for potable water as well as reclaimed water will be gradually increased over a two-year period.
- To help encourage large water-using organisations become more water efficient, the WEF provides co-funding for water efficiency management systems, water audits, water recycling projects, and community-wide water conservation programmes.
- Singapore has implemented a mandatory water efficiency labelling programme that also requires all new developments and existing premises undergoing renovation to install water fittings that have at least a 1-tick water efficiency rating under the labelling programme.
- Singapore requires all large water users to submit WEMPs that break down all water usage, identify areas to reduce water consumption, and detail action plans that save water.
- Singapore's WEB certification scheme encourages non-domestic buildings to put in place water-efficient measures.
- To encourage young people to save water, the Ministry of Education has incorporated water conservation topics in the syllabus while the water utility's school programme involves students becoming junior water advocates.

Energy

- To encourage existing buildings to adopt energy efficiency as well as other resource efficiency–related improvements and measures Singapore offers small and medium business building owners or tenants funding to conduct building retrofits.
- Singapore is providing financial support for environmentally sustainable design consultants to work collaboratively with building designers to ensure energy efficiency measures are incorporated early in the design stage. This assistance involves the consultants holding workshops as well as assisting with energy simulation studies. Meanwhile, funding is provided for companies to carry out energy appraisals in existing facilities.

- To encourage investors to incorporate energy and resource efficiency practices in their new or renovated facilities, Singapore provides co-funding for Design for Efficiency workshops that involve external and internal experts identifying energy and resource efficiency opportunities.
- Singapore provides productivity grants that reduce energy usage and improve efficiency for industrial facilities that have implemented an energy management system.
- To encourage the uptake of public transportation, Singapore has developed an ERP tool that charges motorists for using congested stretches of the road.
- Singapore's green building innovations cluster enables local companies to test novel technologies and gather data, with the eventual aim of commercialising the technology.
- Singapore's energy efficiency partnership programme aids companies that wish to be more energy efficient by providing them with a variety of resources including learning network activities.

Food

- To enhance agricultural productivity, Singapore is providing funding for upgrading of facilities, enhancing productivity through installation of automated systems and for research and development.
- Singapore has implemented a range of policies that intensifies land use, including ensuring nearly all farmland is used for production, setting minimum production levels, and requiring renewal of land tenures periodically.
- Singapore provides technical expertise to food companies to enhance productivity and reduce waste.

Overall, Singapore uses a variety of fiscal and non-fiscal tools to create interdependencies and synergies between the nexus systems while reducing trade-offs between the systems in the development of a green economy. These tools are summarised in Table 5.10.

Table 5.10 Singapore case summary

Tool	Tool type	Policy title	Description	WEF sectors addressed
Fiscal	Market-based instruments and pricing	Potable and NEWater Price revisions	A range of water infrastructure investments and rising operational costs have made the revision of water prices necessary.	Water
		Electronic Road Pricing	This pay-as-you-use tool charges motorists for using congested stretches of the road.	Energy
	Financial incentives	Green Mark Incentive Scheme for existing buildings and premises	Encourages building owners, occupants, and tenants to undertake and adopt energy-efficient improvements and measures within their buildings and premises.	Energy
		Water Efficiency Fund	Encourages large water-using organisations to seek efficient ways to manage their water demands.	Water
		Green Mark Incentive Scheme—Design Prototype	Encourages developers and building owners to include energy efficiency in the buildings' design stage.	Energy
		Energy Efficiency Improvement Assistance Scheme	Provides assistance for companies to carry out energy appraisals and identify potential areas for energy efficiency improvements in their facilities.	Energy
		Design for Efficiency Scheme	Encourages investors in new facilities and facility expansion projects to incorporate energy and resource efficiency practices early in the design stage.	Energy, water
		Grant for Energy Efficient Technologies	Encourages owners and operators of new and existing industrial facilities to invest in energy-efficient equipment or technologies.	Energy
		Agricultural Productivity Fund	Supports equipment upgrades, adoption of hi-tech systems, and R&D that enhances productivity	Food

(continued)

Table 5.10 (continued)

Tool	Tool type	Policy title	Description	WEF sectors addressed
Non-fiscal	Regulations	Achieving minimum environmental standards in existing buildings	Minimum Green Mark Standards for existing buildings must be met when installing or retrofitting a cooling system; carrying out energy audits on the cooling system; and complying with design system efficiency standards.	Energy
		Water Efficiency Management Plan	All large water users are required to submit a plan that includes actions to save water.	Water
		Mandatory energy management requirements	Energy-intensive companies must appoint an energy manager, report energy use, and submit energy efficiency improvement plans.	Energy
		Intensifying land use	Local farms must ensure maximum use of farmland for production and achieve minimum production levels.	Food
	Standards and mandatory labelling	Mandatory Water Efficiency Labelling Scheme	Suppliers of a range of products are required to label the water efficiency of their water fittings and appliances on all displays, packaging, and advertisements.	Water
	Public education and skills development	Training green specialists	The BCA Academy is training green building professionals, specialists, and experts in the design and management of sustainable buildings.	Energy, water
	Public-private partnerships	10% challenge	PUB challenges non-domestic water users to save water, and provides tips on how to do so.	Water
	Stakeholder participation	Carbon footprint assessment tool	BCA has been working with non-governmental organisations and researchers to develop a carbon footprint assessment toolkit for buildings.	Energy
		Enhancing indoor environmental quality	BCA is working with industry to review various building codes to ensure good indoor environmental quality.	Energy

(continued)

Table 5.10 (continued)

Tool	Tool type	Policy title	Description	WEF sectors addressed
	Information and awareness-raising	Water Efficient Building design guide	The guide provides guidelines and recommendations on how industry can be water-efficient.	Water
		Water Efficient Homes programme	Residents are encouraged to save water and install water-efficient devices. This is bolstered by the utility visiting households.	Water
	School education	Green School Roadmap	The roadmap gives younger generations first-hand experience in sustainability issues.	Energy
		Water conservation education	Water conservation topics are included in the social studies syllabus and students are given a water conservation guide for around the house.	Water
	Voluntary labelling	Green Mark Scheme	A three-tiered rating system that evaluates a building's environmental impact and recognises its sustainability performance. It has now been expanded beyond buildings and includes four main categories of climate, energy, resource use, and smart and healthy buildings.	Water, energy, food
		Water Efficient Building certification	Encourages non-domestic buildings to put in place water-efficient measures.	Water
	Clustering policies	Green Building Innovation Cluster	The cluster aims to increase the deployment of energy-efficient building solutions.	Energy
		Food Innovation Resource Centre	The centre helps food companies optimise productivity through the adoption of technology, process re-engineering, automation, and design customisation.	Food

(continued)

Table 5.10 (continued)

Tool	Tool type	Policy title	Description	WEF sectors addressed
	Demonstration projects	HDB Greenprint Fund	Funds and supports test-bedding of smart and sustainable solutions.	Water, energy, food
		HDB's Eco-Precinct	Enables the testing of potential/workable smart and sustainable solutions.	Water, energy, food
		Green Building Innovation Cluster energy-efficient demonstrations scheme	Serves to demonstrate energy-efficient technologies in buildings to generate local performance data for verification, from which successful technologies will be replicated in other buildings and eventually commercialised.	Energy
	Awards and public recognition	Green Mark Pearl award	Recognises developers and others for outstanding achievements in creating environmental sustainability in their project/building.	Water, energy, food
	Voluntary agreements	Energy Efficiency National Partnership	The voluntary partnership programme is for companies that wish to be more energy efficient, with participants provided advice and support.	Energy

Notes

1. Monetary Authority of Singapore. 2017. *The Singapore economy* [Online]. Available: http://www.sgs.gov.sg/The-SGS-Market/The-Singapore-Economy.aspx.
2. Monetary Authority of Singapore. 2016. The Singapore economy. *Macroeconomic review, October 2016* [Online]. Available: http://www.mas.gov.sg/~/media/resource/publications/macro_review/2016/Oct%2016/Chapter%202_Oct16.pdf.
3. National Population and Talent Division. 2016. Population in brief 2016. Available: https://www.nptd.gov.sg/Portals/0/Homepage/Highlights/population-in-brief-2016.pdf.
4. PUB. 2016b. Our water, our future. Available: https://www.pub.gov.sg/Documents/PUBOurWaterOurFuture.pdf.
5. PUB. 2017d. *Singapore water story* [Online]. Available: https://www.pub.gov.sg/watersupply/singaporewaterstory.
6. PUB. 2016b. Our water, our future. Available: https://www.pub.gov.sg/Documents/PUBOurWaterOurFuture.pdf.
7. Energy Market Authority. 2016. Singapore energy statistics 2016. Available: https://www.ema.gov.sg/cmsmedia/Publications_and_Statistics/Publications/SES/2016/Singapore%20Energy%20Statistics%202016.pdf.
8. Centre for Liveable Cities. 2015. A case study of Singapore's smart governance of food. Available: http://www.clc.gov.sg/documents/books/Smart_food_governance_paper-SG_case_study_FINAL_Sept%2028_3.pdf.
9. Housing and Development Board. 2017. *HDB Greenprint fund* [Online]. Available: http://www.hdb.gov.sg/cs/infoweb/about-us/our-role/smart-and-sustainable-living/hdb-greenprint/funding.
10. Building and Construction Authority. 2017b. *BCA Green mark assessment criteria and application forms* [Online]. Available: https://www.bca.gov.sg/GreenMark/green_mark_applnforms.html.
11. Building and Construction Authority. 2015. Build green.
12. Building and Construction Authority. 2013a. BCA's inaugural Greenovate challenge engages youths to green school buildings. Available: https://www.bca.gov.sg/newsroom/others/pr22032013_GI.pdf.
13. Building and Construction Authority. 2014. 3rd Green building masterplan. Available: https://www.bca.gov.sg/GreenMark/others/3rd_Green_Building_Masterplan.pdf.

14. Ibid.
15. Ibid.
16. PUB. 2017e. *Water efficiency management plan* [Online]. Available: https://www.pub.gov.sg/savewater/atwork/managementplan.
17. PUB. 2016b. Our water, our future. Available: https://www.pub.gov.sg/Documents/PUBOurWaterOurFuture.pdf.
18. PUB. 2017a. *About Water efficiency labelling scheme* [Online]. Available: https://www.pub.gov.sg/wels/about.
19. PUB. 2017e. *Water efficiency management plan* [Online]. Available: https://www.pub.gov.sg/savewater/atwork/managementplan.
20. PUB. 2016c. Water balance chart.
21. PUB. 2016b. Our water, our future. Available: https://www.pub.gov.sg/Documents/PUBOurWaterOurFuture.pdf.
22. Building and Construction Authority. 2014. 3rd Green building masterplan. Available: https://www.bca.gov.sg/GreenMark/others/3rd_Green_Building_Masterplan.pdf.
23. Building and Construction Authority. 2017a. *$5 million Green mark incentive scheme—Design prototype (GMIS-DP)* [Online]. Available: https://www.bca.gov.sg/greenmark/gmisdp.html.
24. International Partnership for Energy Efficiency Cooperation. 2015. Building code implementation—Country summary. Available: http://www.gbpn.org/sites/default/files/Singapore_Country%20Summary_0.pdf.
25. National Environment Agency. 2017. *Design for efficiency scheme* [Online]. Available: http://www.nea.gov.sg/grants-awards/energy-efficiency/design-for-efficiency-scheme-(dfe).
26. Energy Efficient Singapore. 2017b. Productivity grant (Energy efficiency) formally known as Grant for energy efficient technologies (GREET).
27. SUN, S.G. and SAM, P.Y. (2015). Sustainable Transport in Singapore. https://www.lta.gov.sg/ltaacademy/pdf/J15May_p30Sun_SustainableTransportInSingapore.pdf
28. Building and Construction Authority. 2017c. *Green building innovation cluster* [Online]. Available: https://www.bca.gov.sg/ResearchInnovation/gbic.html.
29. Energy Efficient Singapore. 2017a. *Energy efficiency national partnership (EENP)* [Online]. Available: http://www.e2singapore.gov.sg/Programmes/Energy_Efficiency_National_Partnership.aspx.
30. Ibid.
31. Food Innovation and Resource Centre. 2017. *Process engineering innovation* [Online]. Available: http://www.firc.com.sg/food-automation-unit/.

References

Agri-Food and Veterinary Authority of Singapore. 2016. Agriculture productivity fund and agriculture policies for local farms. http://app.mnd.gov.sg/Portals/0/Annex%20A_APF_AVA.pdf

Building and Construction Authority. 2013a. BCA's inaugural Greenovate challenge engages youths to green school buildings. https://www.bca.gov.sg/newsroom/others/pr22032013_GI.pdf

———. 2013b. Singapore leading the way for green buildings in the tropics. https://www.bca.gov.sg/greenmark/others/sg_green_buildings_tropics.pdf

———. 2014. 3rd green building masterplan. https://www.bca.gov.sg/GreenMark/others/3rd_Green_Building_Masterplan.pdf

———. 2015. Build green.

———. 2017a. *$5 million Green Mark incentive scheme—Design prototype (GMIS-DP)* [Online]. https://www.bca.gov.sg/greenmark/gmisdp.html

———. 2017b. *BCA Green Mark assessment criteria and application forms* [Online]. https://www.bca.gov.sg/GreenMark/green_mark_applnforms.html

———. 2017c. *Green building innovation cluster* [Online]. https://www.bca.gov.sg/ResearchInnovation/gbic.html

Centre for Liveable Cities. 2015. A case study of Singapore's smart governance of food. http://www.clc.gov.sg/documents/books/Smart_food_governance_paper-SG_case_study_FINAL_Sept%2028_3.pdf

Energy Efficient Singapore. 2017a. *Energy efficiency national partnership (EENP)* [Online]. http://www.e2singapore.gov.sg/Programmes/Energy_Efficiency_National_Partnership.aspx

———. 2017b. Productivity grant (energy efficiency) formally known as Grant for energy efficient technologies (GREET).

Energy Market Authority. 2016. Singapore energy statistics 2016. https://www.ema.gov.sg/cmsmedia/Publications_and_Statistics/Publications/SES/2016/Singapore%20Energy%20Statistics%202016.pdf

Food Innovation and Resource Centre. 2017. *Process engineering innovation* [Online]. http://www.firc.com.sg/food-automation-unit/

Housing and Development Board. 2017. *HDB Greenprint fund* [Online]. http://www.hdb.gov.sg/cs/infoweb/about-us/our-role/smart-and-sustainable-living/hdb-greenprint/funding

International Partnership for Energy Efficiency Cooperation. 2015. Building code implementation—Country summary. http://www.gbpn.org/sites/default/files/Singapore_Country%20Summary_0.pdf

Monetary Authority of Singapore. 2016, October. The Singapore economy. *Macroeconomic Review* [Online]. http://www.mas.gov.sg/~/media/resource/publications/macro_review/2016/Oct%2016/Chapter%202_Oct16.pdf

——. 2017. *The Singapore economy* [Online]. http://www.sgs.gov.sg/The-SGS-Market/The-Singapore-Economy.aspx

National Environment Agency. 2017. *Design for efficiency scheme* [Online]. http://www.nea.gov.sg/grants-awards/energy-efficiency/design-for-efficiency-scheme-(dfe)

National Population and Talent Division. 2016. Population in brief 2016. https://www.nptd.gov.sg/Portals/0/Homepage/Highlights/population-in-brief-2016.pdf

PUB. 2016a. Application for Water Efficiency Fund (WEF). https://www.pub.gov.sg/Documents/WEF_ApplicationForm_2016.pdf

——. 2016b. Our water, our future. https://www.pub.gov.sg/Documents/PUBOurWaterOurFuture.pdf

——. 2016c. Water balance chart.

——. 2017a. *About Water efficiency labelling scheme* [Online]. https://www.pub.gov.sg/wels/about

——. 2017b. *Efficiency measures* [Online]. https://www.pub.gov.sg/savewater/atwork/efficiencymeasures

——. 2017c. *Industrial water prices* [Online]. https://www.pub.gov.sg/PublishingImages/T5A%20NEWater.jpg

——. 2017d. *Singapore water story* [Online]. https://www.pub.gov.sg/water-supply/singaporewaterstory

——. 2017e. *Water efficiency management plan* [Online]. https://www.pub.gov.sg/savewater/atwork/managementplan

——. 2017f. *Water efficient building certification programme* [Online]. https://www.pub.gov.sg/savewater/atwork/certificationprogramme

——. 2017g. *Water prices* [Online]. https://www.pub.gov.sg/PublishingImages/T1A%20dom.jpg

Yew, G. S. A. S. P. 2015. Sustainable transport in Singapore. *Journeys.*

6

The Green Economy and the Water-Energy-Food Nexus in Massachusetts

Introduction

Massachusetts is one of the most densely populated states in the United States and home to nearly half of all New England residents. The population of over 6.7 million is concentrated around Boston. Half of the state's land is forested and the region's longest river, the Connecticut, with its fertile river valley, cuts across the state, with 12 hydropower dams harnessing its energy. In 2015, Massachusetts's GDP was $484.9 billion and ranked 11th in the United States. In 2015, the state's real GDP grew by 3.8 percent.[1] The main economic sectors of Massachusetts include financial services and real estate; professional, technical, and scientific services; information technology; health care; and computer and electronic products manufacturing.[2]

6.1 Water-Energy-Food Nexus Pressures

Massachusetts is experiencing a variety of water, energy, and food nexus pressures that are detrimental to the development of a green economy as described below through a variety of examples.

© The Author(s) 2018
R.C. Brears, *The Green Economy and the Water-Energy-Food Nexus*,
DOI 10.1057/978-1-137-58365-9_6

Water

On average, Massachusetts has one of the lowest per capita residential water demands in the United States. Massachusetts's economy is linked to its natural resources, particularly water. Despite the state receiving on average 45 inches of rainfall each year, rainfall can vary significantly from year to year and can drop to below 30 inches during a severe drought year. Short-term droughts can also severely deplete water supplies as well as source rivers, streams, and ponds. Also, weather patterns are seasonal with evapotranspiration increasing in summer, resulting in less rainfall available to contribute to recharge. Massachusetts is also one of the most densely populated states with over six million people living on around six million acres of land. As such, the per capita water availability is significantly less than some desert states such as Nevada. Therefore, Massachusetts's current water use and future growth are potentially constrained by variations in the availability of water resources.[3]

Energy

Massachusetts has long been dependent on imported fossil fuels to generate its electricity with 64 percent of its electricity generated from natural gas and 7 percent of its electricity from coal in 2015.[4] Meanwhile, over 27 percent of Massachusetts's residents use fuel oil as their primary heating fuel, which is more than five times the national average.[5] This reliance on fossil fuels for electricity generation is reflected in Massachusetts's electricity prices being higher than the US average.[6] Furthermore, while natural gas prices are low for now, they are historically volatile with wide price swings tied to increasing demand, extreme weather events, and uncertainties about available gas supplies. As such, price swings are often felt most by electricity consumers in states, such as Massachusetts, that rely more heavily on natural gas.[7] In 2015, almost 10 percent of net electricity generation came from renewable energy resources, with nearly two-thirds from solar, wind, and biomass and the remaining one-third from hydroelectricity.

Food

There are over 7000 farms in Massachusetts working on over 500,000 acres to produce $492 million in agricultural products.[8] Massachusetts is one of the largest growers of cranberries in the nation. The rest of its agricultural production includes nurseries, fruit orchards, and vegetable production. Agricultural production is challenged by a short growing season along with limited farmland availability: the state is one of the most densely populated in the United States resulting in farmland making up only one-tenth of the state's land, of which half of the farmland is woodland or pasture.[9,10] In addition, future agricultural production is challenged by the fact that 80 percent of the state's farms are family owned, 95 percent of farms are considered 'small farms' with sales below $250,000, and the average age of a Massachusetts principal operator is 57.8 years.[11]

6.2   Water: Fiscal Tools

Massachusetts has implemented a variety of water-related fiscal tools that create interdependencies and synergies between the nexus systems while reducing trade-offs between the systems in the development of a green economy.

Water Quality Management Funding

In 2017, the Massachusetts Department of Environmental Protection (MassDEP), Bureau of Water Resources is distributing grant funds towards projects that identify water quality problems and provide preliminary designs for best management practices. The goal of the grant funding is to:

- Determine the nature, extent, and causes of water quality problems
- Develop municipal and regional approaches to stormwater issues
- Assess the impacts and determine pollutant load reductions that improve water quality

- Develop green infrastructure projects that manage wet weather to maintain or restore natural hydrology
- Develop preliminary designs and implement plans that address water quality impairments in impaired watersheds

With funding of up to $180,000 per project the following types of assessment/planning tasks that focus on watershed or sub-watershed-based non-point source pollution will be considered for funding:

- Water quality assessments that identify and characterise specific non-point source pollution problem sites
- Development of stormwater utilities in regulated and non-regulated communities
- Green infrastructure and low-impact development assessments and conceptual designs for projects that manage wet weather
- Development of implementation plans including conceptual drawings and engineering studies that lead to remediation of water quality and restoration for beneficial uses[12]

Water Innovation Trust Initiatives

In 2015, Massachusetts launched the Water Innovation Trust with $800,000 in initial funding available for innovative water projects throughout the state. In partnership with the MassDEP, Massachusetts Clean Energy Center (MassCEC) will provide support to the development of test-bed networks and assist municipal wastewater treatment plants in adopting energy efficiency and innovative water treatment technologies. Grants will be awarded for:

- Feasibility studies at existing and potential test-bed locations that can accelerate the commercialisation of cutting-edge water technologies
- The development of innovative wastewater treatment technologies that reduce electricity consumption, cut energy costs for communities, and/or improve the treatment process[13]

Water Investment Support

MassCEC provides a range of funding opportunities that support the development, demonstration and commercialisation of innovative water technologies.

The Catalyst Programme

This programme provides grants of up to $65,000 to researchers and early-stage companies demonstrating initial prototypes of their water technologies.[14]

The InnovateMass Programme

The programme provides up to $150,000 in grant funding and technical support to applicant teams demonstrating commercial readiness of innovative technologies at demonstration sites.[15]

6.3   Water: Non-fiscal Tools

Massachusetts has implemented a variety of water-related non-fiscal tools that create interdependencies and synergies between the nexus systems while reducing trade-offs between the systems in the development of a green economy.

DeployMass Programme

DeployMass, formerly known as the Massachusetts as a First Customer Program, facilitates the adoption of water innovation technologies at public agencies, public academic institutions, and municipalities across the Commonwealth to support the growth and development of Massachusetts-based companies while saving taxpayer dollars.

Water Technology Industry Roadmap

MassCEC, in collaboration with Battelle Memorial Institute and Redwood Innovation Partners, published the 2015 Water Technology Industry Roadmap that describes the rapidly developing water technology industry in Massachusetts and identifies strategies to increase the impact of the industry in the overall Massachusetts economy.[16]

New England Water Innovation Network

MassCEC has played a key role in the creation of the New England Water Innovation Network that comprises a diverse network of innovators, students, engineering firms, researchers, equipment providers, test-beds, and government officials brought together to accelerate the introduction and commercialisation of innovative water technologies.[17]

Water Contests for the 2016–2017 School Year

The Massachusetts Water Resource Authority (MWRA) sponsors the Poster, Writing and Video Contests for students who live in or attend a school in a community served by the MWRA. In 2016–2017 the topic is 'Whatever the Weather' with contests in the following categories:

- *Poster contest:* Students are to create a poster that shows how weather affects the MWRA and its workers, with posters judged in three category grades: K–2, Grades 3–5, and Grades 6–8
- *Writing contest:* Students are to write an essay or poem to show how weather affects the MWRA and its workers. Entries will be judged in two categories: Grades 3–5 and Grades 6–8
- *Video contest:* Students are to produce a video to show how weather affects the MWRA and its workers

Prizes and Recognition

The winners in each category each receive a $100 gift card, second prize is a $50 gift card, and third prize is a $25 card, with all winners, including honourable mention winners, invited with their families to an Awards Ceremony following the contest.[18]

6.4   Energy: Fiscal Tools

Massachusetts has implemented a variety of energy-related fiscal tools that create interdependencies and synergies between the nexus systems while reducing trade-offs between the systems in the development of a green economy.

Massachusetts Electric Vehicle Incentive Program: Workplace Charging

The Massachusetts Electric Vehicle Incentive Program (MassEVIP): Workplace charging is an open, first-come-first-served grant programme administered by the MassDEP that provides incentives for employers to acquire Level 1 (standard wall outlet that provides 120 volts, typically for overnight charging of 8–20 hours) and Level 2 (a free-standing or hanging charger station that provides 240 volts of charging power for quicker charges of four to eight hours) electric vehicle (EV) charging stations that can charge EVs produced by multiple manufacturers. Under the programme, employers that employ more than 15 or more persons in a non-residential place of business in the state can receive 50 percent of the funding up to $25,000 for hardware costs to employers for installation of Level 1 and Level 2 charging stations. If the employer has multiple sites across Massachusetts the employer is eligible to apply separately for each location.[19]

Pathways to Zero Net Energy Buildings

In 2008, the governor of Massachusetts issued a challenge to the building industry in the state to transform the building sector and help put Massachusetts on a path towards zero net energy buildings (ZNEB). A ZNEB is one that is optimally efficient and over a course of a year generates energy on-site using clean renewable resources in a quantity equal to or greater than the total amount of energy consumed on-site. To support the emerging market of ZNEB practices in Massachusetts, the Department of Energy Resources (DOER) has established the Pathways to Zero Net Energy Program that aims to facilitate scalable and dramatic improvements in the energy performance of new and existing buildings in the Commonwealth through the design and construction of ZNEBs.

ZNEB Funding Opportunities

The programme is providing grant funding for existing building renovations and new construction projects that demonstrate a goal of ZNEB energy performance. With $2.5 million in funding available the programme will award projects with funding ranging from $10,000 to $500,000. DOER anticipates funding around 30–50 residential projects and 1–3 commercial and institutional projects.[20]

The Commonwealth Wind Incentive Program

The Commonwealth Wind Incentive Program provides rebate and grant funding for community and commercial wind projects throughout Massachusetts. The goal of the incentive programme is to, first, guide and support wind project applicants through the detailed analysis and community engagement necessary to evaluate proposed projects, and second, support the development of projects that, through the results of step one, demonstrate appropriate siting. The programme offers community and commercial wind projects:

- Site assessment grants for services that provide a preliminary evaluation of the potential for wind development on one or more sites
- Feasibility study grants to conduct an in-depth analysis of the technical, environmental, regulatory, and financial aspects of a wind project at a specific site
- Development grants for development activities including permitting, environmental impact evaluation, geotechnical studies, public outreach[21]

Organics-to-Energy Grant Programme

MassCEC's Commonwealth Organics-to-Energy grant programme supports the use of anaerobic digestion and other technologies to convert source-separated organic wastes into electricity and thermal energy. There are three separate funds that can be awarded under this programme, which are summarised in Table 6.1.[22]

Table 6.1 MassCEC's organics-to-energy grant programme

Grant	Description	Grant amount
Technical services/ technical study grants (only public entities can apply)	Activities eligible for the grant include: technical assistance in the development, evaluation, and procurement of contracts; technical assistance for proposals to site organics-to-energy-facilities; public engagement processes for matching community needs with organics processing options; and pre-feasibility studies for sites, generator clusters, or technical approaches to handling identified organic waste streams	Up to $50,000 is available per grant and a 10 percent cost share is required

(continued)

Table 6.1 (continued)

Grant	Description	Grant amount
Feasibility studies that both public and private (profit or not-for-profit) entities can apply for	Activities that are supported by grant funding include assessing feedstock, the technical and engineering feasibility of the project, and identifying any community impacts or issues.	Grants of up to $40,000 are available with a 5 percent cost share for public entities and 20 percent cost share for non-public entities
Implementation and pilot project funding are available for both public and private (profit or not-for-profit)	Grants available for entities engaging in activities including designing, permitting and constructing as well as installing and/or commissioning of equipment.	Implementation projects may be eligible for 75 percent of design phase costs and 25 percent of construction phase costs up to $400,000 and pilot projects may be eligible for 75 percent of design phase costs and 50 percent of construction phase costs up to $200,000

6.5   Energy: Non-fiscal Tools

Massachusetts has implemented a variety of energy-related non-fiscal tools that create interdependencies and synergies between the nexus systems while reducing trade-offs between the systems in the development of a green economy.

Advanced Building Energy Codes

In 2008, Massachusetts adopted the requirement that building energy codes meet or exceed the latest International Energy Conservation Code and stay current with its three-year update cycle. In addition, the Commonwealth developed a 'stretch' energy code. The stretch code is an optional appendix to the Massachusetts Building Energy Code that allows cities and towns to choose a more energy-efficient option. This option increases the efficiency requirements in any municipality that adopts it, for all new residential and many new commercial buildings as well as residential additions and renovations that would normally trigger building code energy efficiency requirements.[23] The stretch code, unlike traditional codes that prescribe specific energy requirements for new building components, is a performance-orientated code that mandates a percentage reduction in total building energy use, while allowing developers to make their own design choices on how to achieve this reduction.[24]

Building Energy Rating and Labelling

Between 2012 and 2014 DOER implemented a pilot building energy rating and labelling programme for both residential and commercial buildings that enables comparisons between buildings with the energy rating of the building based on its physical characteristics including the level of insulation, efficiency of the HVAC system, independent of tenant or user behaviour. The 'MPG rating' shows the home's energy performance score (the estimated total energy use of the home in one year) as well as the home's carbon footprint (the estimated carbon emissions based on the annual amounts, types, and sources of fuels used): the lower the score of both, the better. The scorecard presents these metrics for the home in its current state as well as the expected metrics if recommended efficiency improvements were made and a comparison to the average home in the pilot communities.[25]

Commonwealth Building Energy Intelligence Programme

In 2016, the Baker-Polito Administration announced it was signing a $5.6 million three-year deal with a Boston company to use advanced energy metering and analytics to identify opportunities for increased efficiency at state properties. The Commonwealth Building Energy Intelligence Program will track in real time energy usage in state buildings and provide recommendations for reducing energy use. Specifically, the programme will identify energy anomalies as they occur, prioritise energy projects that target under-performing buildings, and identify billing errors on utility bills. Facilities that will come under the programme include state office buildings, residence and dining halls at state universities, administration and classroom buildings at community colleges, state hospitals, and correction facilities.[26]

MassCEC Green Workforce Development Programmes

In 2009, MassCEC and Commonwealth Corporation awarded six grants totalling $1 million to develop comprehensive workforce development programmes for the clean energy industry. The grantees produced more than 30 lesson plans, certificates and training programmes for high school and college students, at-risk youth and low-income populations, building and trade professionals, and clean energy employees. The topics covered include solar PV installation, energy efficiency practices, and best practices in developing train-the-trainer programmes for vocational schools as well as universities that teach engineering, business, politics, and economics of clean energy.[27]

Clean Energy Internship Program

MassCEC offers the Clean Energy Internship Program which is designed to help train a qualified clean energy workforce. Since 2011, the programme has placed over 1000 students in internships with 280 different clean energy employers across the state. The internship programme provides valuable support in particular to small businesses of all sectors, with the majority of interns working at renewable energy companies and energy efficiency companies.[28]

Learn and Earn Programme

As part of the effort to build a clean energy workforce MassCEC has launched the Learn and Earn programme for high school students to be employed and receive training in advanced clean energy curriculum.[29] Students will be provided career exploration, work readiness training, paid work–based learning that focuses on clean energy, and dual enrolment that provides credit from a high school education institution.[30]

Successful Women in Clean Energy Programme

In 2015, MassCEC launched the Successful Women in Clean Energy programme to prepare low- and moderate-income women for clean energy positions in sales and business management. The individuals trained will be placed into six-month paid fellowships at Massachusetts clean energy companies.[31]

MassCECs Cleanweb Hackathon

In 2015, MassCEC held its fourth annual Boston Cleanweb hackathon that brought together more than 70 innovators from the energy, technology, and business sectors to create 16 new clean energy mobile and online technologies. Participants had just one week to create their web-based energy apps before pitching to a panel of judges, who awarded more than $11,000 in prizes. Participating teams were also able to participate in the Cleanweb Haccelerator, an eight-week business accelerator programme to continue their project development and compete for an additional $5000.[32]

6.6   Food: Fiscal Tools

Massachusetts has implemented a variety of food-related fiscal tools that create interdependencies and synergies between the nexus systems while reducing trade-offs between the systems in the development of a green economy.

Agricultural Preservation Restriction Program

The Agricultural Preservation Restriction Program is a voluntary programme that offers a non-development alternative to farmland owners for their agricultural lands who are faced with a decision regarding their future use. The programme offers farmers a payment up to the difference between the 'fair market value' and the 'fair market agricultural value' of their farmland in exchange for a permanent deed restriction that prevents any use of the property that will have a negative impact on its agricultural viability. To be eligible for the programme farms must be at least five acres in size, land must have been devoted to agriculture for at least two years and made at least $500 in gross sales per year for the first five acres, plus $5 for each additional acre, or 50 cents per additional acre of woodland and/or wetland.[33,34]

Agricultural Preservation Restriction Improvement Program

The Agricultural Preservation Restriction Improvement Program is designed to help sustain active commercial farming on land that has been protected by an agricultural preservation restriction. The programme provides business planning and technical assistance to eligible farmers in order to improve farm productivity and profitability. A team of consultants help each farm achieve long-term agricultural use and viability. One of the following objectives needs to be identified for a participant to receive funding to implement business plans:

- Improve the economic viability of the farm
- Retain or create private sector jobs and tax revenue directly or indirectly
- Improve farm productivity and competitiveness
- Expand farm facilities as part of a modernisation or business plan
- Support renewable energy or environmental projects[35]

Massachusetts Farm Energy Program

The Massachusetts Farm Energy Program (MFEP) is a joint project of the Center for EcoTechnology and the Massachusetts Department of Agricultural Resources (MDAR). MFEP offers a range of services to the farming community to reduce their energy use and produce renewable energy.

MFEP Incentives for Energy Efficiency

MFEP offers financial incentives for implementing energy-saving upgrades. Farmers may be eligible to receive up to $5000 towards the costs of energy efficiency retrofits. The intention is to help farmers' complete upgrades recommended by public utilities or custom energy assessments.[36]

MFEP Custom Audits

MFEP provides customised farm energy audits to help farmers assess energy savings and renewable energy costs. MFEP pays for 75 percent of the costs. To manage costs audits are targeted to address specific concerns and measures identified in the application process that have not been adequately addressed by other audit programmes.[37]

MFEP Renewable Energy Assessments

MFEP provides funding up to 75 percent of the cost of an assessment or consultation for renewable energy system information that is not readily accessible through installers with a high priority placed on assessments for solar thermal, biomass for solid fuels, anaerobic waste digesters, wind, photovoltaics, geothermal, and hydro.[38]

Agricultural Energy Grant Program

The Agricultural Energy Grant Program, run by the MDAR, provides grant funding for agricultural projects that improve energy efficiency and facilitate the adoption of alternative clean energy technologies. The programme distributes grants up to a maximum of $30,000 per applicant. Projects eligible for grant funding include:

- Dairy energy efficiency improvements
- Greenhouse and nursery heating and ventilation improvements
- Maple sugaring energy efficiency improvements

- Other technologies that, for example, recover heat, produce efficient heating or cooling
- Renewables including solar photovoltaic, wind, thermal, biofuel production[39]

Agricultural Energy 'Special Projects' Grant Programme

The MDAR provides funding for agricultural energy projects that are geared towards specific, higher capital cost, energy-saving and energy replacement technology implementation opportunities that improve energy efficiency and facilitate alternative clean energy needs within agricultural operations. The programme's total funding is $350,000 with maximum funding per applicant dependent on the category of the special project:

- Energy efficiency projects including heat recovery for anaerobic digesters and efficient walk-in coolers can receive a maximum grant of $25,000.
- Renewable energy projects involving dual use of land solar photovoltaics, which enable the utilisation of farm land for crop production and energy generation, can receive funding up to $100,000.
- Energy efficiency/renewable energy projects involving zero net energy greenhouse projects can receive a maximum grant of $75,000; super-efficient new building construction for on-site packing and storage and other relevant agricultural structures can receive a maximum grant of $75,000; and commercial-scale, high-efficiency, and renewable energy urban agricultural greenhouses that reside in urban areas can receive a maximum grant of $100,000.[40]

Farm Viability Enhancement Program

The Farm Viability Enhancement Program is designed to diversify and modernise an existing farm's operations. The programme involves selected farmers having a team of professionals from various disciplines including agriculture, marketing, finance, management, and environmental science assembled for each farm to make farm-specific recommendations for improving the viability of the farm. The team provides farmers with technical assistance and the development of a business plan at no cost to farmers.

Farmers who are willing to sign an agricultural covenant that prohibits any land use on the farm except agricultural use are eligible to receive the following funding to implement the recommendations:

- Up to $25,000 is available for farmers willing to agree to a covenant for a period of five years.
- Up to $50,0000 for farmers willing to agree to a ten-year covenant.
- $75,000 or more to farmers meeting certain additional acreage and production thresholds and agreeing to a ten-year covenant.

The application is by competitive process in which applicants are evaluated and selected on the basis of degree of threat to the continuation of agriculture on the land, the number of acres to be placed in the programme, the current intensity of use on the farm and its significant contribution to the state's agricultural industry, whether the farm has diversified into retail or value-added activities, the experience of the operators, whether environmental objectives would be accomplished through the programme, and the farm's productivity of the land.[41]

Urban Agricultural Grant Program

The MDAR is seeking proposals from municipalities, non-profit organisations, state agencies, public or non-profit educational or public health institutions and established urban farms for projects that will advance commercial urban food production in the Commonwealth. Project proposals must focus on infrastructure improvements, technical assistance, building upgrades, land procurement, and purchase of farm equipment in communities with populations of at least 35,000 or more, a median annual household income below the state average, and/or be located within neighbourhoods designated as food deserts by the US Department of Agriculture. Examples of priority projects include:

• Soil management initiatives that improve soil quality in urban environments
• Land acquisition proposals for food production in urban settings
• Marketing, distribution, and transport initiatives that improve the transportation and distribution of locally produced grown products from farm to customer
• Green infrastructure projects that help urban farmers scale up the volume and quality, enable year-round production or to manage energy and water usage, or allow for more intensive and efficient food production in urban environments
• Innovative growing technology that demonstrates practical/economically viable approaches to urban aquaculture/aquaponics and vertical farming and other innovative growing methods
• Urban to rural bridge projects that strengthen connections between rural and urban agriculture including innovative marketing models, technology, job creation, and food production benefiting and sited within urban neighbourhoods

Grant Amount

Successful applications will receive grants ranging between $5000 and $40,000 (but may exceed this amount at the discretion of MDAR). Funding will also be available for municipalities and non-profit organisations for urban land acquisition proposals via matching grants up to $75,000.[42]

Agricultural Environmental Enhancement Program: Water

The Agricultural Environmental Enhancement Program run by MDAR provides grant funding to support the mitigation and/or prevention of negative impacts to the Commonwealth's natural resources that may result from agricultural practices. The funding, in the form of a reimbursement, is for agricultural operations that implement eligible projects that prevent, reduce, or eliminate environmental impacts. The maximum funding per applicant is capped at $25,000 or 85 percent of total project costs. Examples of eligible projects include:

- Drip irrigation
- Wells and irrigation pumps for use with micro-irrigation systems
- Automated irrigation systems
- Water re-use projects[43]

Massachusetts Food Ventures Program

The Massachusetts Food Ventures Program aims to increase access to healthy, affordable food options and improve economic opportunities for low- to moderate-income communities, as well as support the objectives of the Massachusetts Local Food Action Plan. The programme will provide funding through grants to support food ventures, sited primarily in or near communities of low to moderate income, including rural communities. Reimbursement grants of up to $250,000 will be awarded, with a minimum of 50 percent of the proposed cost that must be matched by cash, for projects that include:

- *Food processing infrastructure*: To ensure the availability of Massachusetts-produced foods that can benefit low- and moderate-income households
- *Non-retail: Food commissaries*: Innovative proposals that connect local farms and partners to create new local food retail markets or improve marketing within low-income communities
- *Food co-ops*: Proposals for construction, renovation, or build-out of existing facilities
- *Greenhouses and farmers' markets*: Develop and build facilities designed to increase year-round access to diverse food production and distribution to existing or new markets to benefit low- to moderate-income communities

- *Food hubs*: Capital for infrastructure to build or renovate sites for new or existing ventures that improve commercial food access in low- to moderate-income communities
- *Commercial markets*: Innovative projects that connect local food producers and partners to develop, build, or renovate a large-scale retail market to improve the commercial distribution of healthy, locally produced food
- *Mobile markets/innovative markets*: To purchase, design, and upgrade vehicles to provide healthy Massachusetts food[44]

6.7   Food: Non-fiscal Tools

Massachusetts has implemented a variety of food-related non-fiscal tools that create interdependencies and synergies between the nexus systems while reducing trade-offs between the systems in the development of a green economy.

Massachusetts Farm Energy Best Management Practices

The Massachusetts farm energy best management practices guide provides the Commonwealth's agricultural community with resources and methods to reduce energy use and produce renewable energy on farms. The best management practices guide recommends on-farm energy upgrades to improve farm viability and minimise the environmental impact of the agricultural industry in Massachusetts by reducing energy consumption, operating costs, emissions, and dependence on fossil fuels. The guide focuses on conventional, cost-effective best management practices for the four primary agricultural sectors in Massachusetts: greenhouses, dairy farms, orchards and vegetable farms and maple sugaring. The guide also provides information on on-farm renewable energy options including wind, solar, thermal, solar photovoltaic, and biomass. Overall, the guide aims to be a practical resource for farmers and service providers to understand farm energy use, evaluate potential equipment upgrades, and prioritise energy efficiency and renewable energy applications.[45]

Commonwealth Quality Programme

Commonwealth Quality is a brand designed by the MDAR that identifies locally sourced products that are grown, harvested, and processed in Massachusetts using practices that are safe and sustainable and do not harm the environment. Commonwealth Quality–certified growers, producers, harvesters, and processors have to not only meet stringent federal, state, and local regulatory requirements for food safety but also employ best management practices and production standards that ensure consumers receive the safest, most wholesome products available.[46] Each sector within the Commonwealth Quality programme, including forestry, aquaculture, produce, lobster, dairy, and maple, has a unique set of requirements to meet in order to use the brand. The requirements are determined through collaboration between industry representatives, regulatory agencies, and research institutes. To become a member of the Commonwealth Quality Program, each prospective member is visited and assessed based on the criteria of their relevant sector.[47]

6.8   Case Study Summary

Massachusetts has implemented a variety of initiatives to create interdependencies and synergies between the nexus systems while reducing trade-offs between the systems in the development of a green economy.

Water

- Massachusetts is providing funding for water quality projects in watersheds that reduce water pollution from non-point sources and stormwater. The funding will be used to conduct assessments, implement stormwater infrastructure as well as develop green infrastructure.

- Massachusetts has implemented a fund that supports the development of water- and wastewater-related technology test-beds that enhance energy efficiency.
- The state provides numerous grants to researchers and early-stage companies to demonstrate water technology prototypes and enhance their commercial readiness.
- To encourage the public sector to procure innovative Massachusetts water technology, the state facilitates the adoption of Massachusetts water technology in public agencies and institutions as well as municipalities.
- To raise awareness of water-related issues among the school-aged population, the state's water resources authority holds a range of water contests for a range of school ages.

Energy

- To encourage uptake of EVs, Massachusetts has initiated the workplace charging funding scheme so employees can charge their vehicles at their non-residential places of business.
- To encourage the building industry to reduce energy use, Massachusetts provides funding for existing building renovations and new construction projects to incorporate renewable energy systems in the building's designs.
- Massachusetts is providing a range of renewable energy grants including an incentive programme for community and commercial wind projects throughout the state as well as for the development of organics-to-energy projects.
- The state has developed a stretch code that is an optional appendix to the state's building energy codes, enabling cities and towns to choose a

more energy-efficient code. This stretch code is performance-based enabling developers to choose the best designs to reduce a building's total energy use.

- Massachusetts has piloted an energy rating and labelling scheme for homes that shows a home's energy performance score as well as its GHG emissions.
- Massachusetts has a range of education and workplace training schemes to create a clean energy workforce including funding certification and training programmes for high school and college students, clean energy internships, and a learn-and-earn scheme for high school students to receive work experience in clean energy. In addition, the state provides training for low-income women in clean energy positions in sales and business management.
- To spark innovation, Massachusetts held a clean-tech haccelerator which drew innovators and start-ups to develop clean energy technologies over a week-long period.

Food

- Massachusetts has an agricultural restriction programme that offers farmers a payment to prevent the use of land for any activities that may impact its agricultural viability. In addition, the state has the Agricultural Preservation Restriction Improvement Program, which helps sustain commercial farming on land that has been protected by the restriction programme with funding available for consultants to increase long-term productivity.
- To increase energy efficiency on farms, Massachusetts offers incentives for energy efficiency retrofits and custom energy audits in addition to renewable energy assessments. Meanwhile, farms can receive grant funding for farms that are making energy efficiency improvements in technological processes as well as implementing renewable energy systems. For high-capital projects, the state has a special grants

programme that enables high-cost energy efficiency projects and alternative energy projects to be undertaken.

- To increase farm productivity, Massachusetts offers a farm viability programme in which a team of experts from various disciplines make recommendations for improving the viability of the farm. As part of this programme, farmers sign a covenant that prohibits the use of any land for activities other than farming-related, with farmers receiving payment for signing the covenant.
- To enhance access to local, healthy food, Massachusetts is currently seeking proposals from a variety of stakeholders to implement commercial urban farming across the state. In addition, the state is funding initiatives that enhance access to healthy food in low- to moderate-income communities with funds available for food-related infrastructure, development of food hubs and farmers' markets, and so forth.
- To reduce the impacts of agricultural production on the state's water resources, Massachusetts provides funds for projects that protect both water quantity and quality.
- To encourage purchasing of environmentally friendly, local food produce, Massachusetts has developed the Commonwealth Quality brand, with sector-specific labelling requirements.

Overall, Massachusetts uses a variety of fiscal and non-fiscal tools to create interdependencies and synergies between the nexus systems while reducing trade-offs between the systems in the development of a green economy. These tools are summarised in Table 6.2.

Table 6.2 Massachusetts case summary

Tool	Tool type	Policy title	Description	WEF sectors addressed
Fiscal	Financial incentives	Water quality management funding	Grants are distributed for projects that identify water quality problems and provide preliminary designs for best practice solutions.	Water, food
		Water Innovative Trust funding	Funding is available for innovative water projects that test new technologies and assist wastewater treatment plants in adopting efficient practices.	Water, energy
		Catalyst programme	Grants for researchers and early-stage companies developing water technologies.	Water
		InnovateMass	Grant funding, and technical support, is available to support innovative technologies at demonstration sites.	Water
		Electric Vehicle charging incentive	Employers can receive a subsidy for installing EV charging stations at workplaces.	Energy
		Zero net energy buildings	Grants are available to new and existing construction projects becoming zero net energy buildings.	Energy
		Commonwealth Wind Incentive Program	Rebates and grant funding are available for community and commercial wind projects with the funding covering site assessments, feasibility studies, and development-related costs.	Energy
		Organics-to-energy grant	Supports the development of organic waste into electricity and thermal energy.	Energy, food
		Agriculture energy grant	Funding for agricultural projects that improve energy efficiency or adopt clean energies.	Energy, food
		Agricultural energy special project grants	Grants are available for high-capital cost, energy-saving, and energy replacement technologies.	Energy, food

(continued)

Table 6.2 (continued)

Tool	Tool type	Policy title	Description	WEF sectors addressed
		Farm viability enhancement	Selected farmers who agree to sign agricultural covenants that prohibit any land use other than for farming receive expert advice on how to diversify and modernise their farms with grants to implement the recommendations.	Food
		Urban agricultural grant	Proposals are being sought by municipalities as well as other stakeholders to advance urban agriculture with successful applicants receiving grants.	Food, water, energy
		Farm energy efficiency incentives	Farmers can receive a grant towards the costs of energy efficiency retrofits.	Energy, food
		Farm custom energy audits	Farmers can receive subsidised energy audits that help assess energy-saving and renewable energy costs.	Energy, food
		Farm renewable energy assessments	Farmers can receive a subsidised assessment or consultation for renewable energy system information that is not readily available through installers.	Energy, food
		Massachusetts Food Ventures Program	Aims to increase access to healthy, affordable food options and improve economic opportunities for low- to moderate-income communities.	Food
	Payment for ecosystem services	Agricultural Preservation Restriction Program	Farmers are offered a payment in exchange for a permanent deed restriction that prevents any land use that will negatively impact the land's agricultural viability. Farmers can also receive advisory services to protect the land and enhance farm productivity.	Food

(continued)

Table 6.2 (continued)

Tool	Tool type	Policy title	Description	WEF sectors addressed
		Agricultural Environmental Enhancement Program	Farmers are reimbursed for agricultural operations that prevent, reduce or eliminate environmental impacts from water.	Food, water
Non-fiscal	Standards and mandatory labelling	Advanced building codes	The state's building energy codes must meet or exceed international standards. Also, the state's stretch code allows cities and towns to choose more energy-efficient codes.	Energy
	Public education and skills development	Green workforce development	The state supports green workforce development programmes that offer courses, certificates, and training programmes.	Energy
		Clean energy internships	The programme helps train a qualified clean energy workforce with interns working in renewable energy and energy efficiency.	Energy
		Learn and earn	High school students are employed and receive training in clean energy.	Energy
		Successful Women in Clean Energy	The programme helps low- and moderate-income women work in clean energy–related sales and business management roles.	Energy
	Information and awareness-raising	Farm energy best management practices	The best management practices guide provides the farming community with resources and methods to reduce energy use and produce renewable energy.	Energy, food
	Clustering policies	New England Water Innovation Network	The state has facilitated the creation of a network to accelerate the development of water technologies.	Water

(continued)

Table 6.2 (continued)

Tool	Tool type	Policy title	Description	WEF sectors addressed
	Voluntary labelling	Building energy rating and labelling	A pilot building rating and labelling programme enabled comparisons between buildings and a home rating showed energy performance.	Energy
		Commonwealth Quality Program	The brand identifies locally sourced products that use safe and sustainable methods.	Food, water
	Resource mapping	Water Technology Industry Roadmap	The map describes the water technology industry and identifies strategies to increase the industry's impact.	Water
		Commonwealth Building Energy Intelligence	The programme will track in real time energy usage in state buildings and provide recommendations for reducing energy use.	Energy
	Public procurement	DeployMass	Facilitates the adoption of water innovation technologies in public facilities.	Water
	Awards and public recognition	Water contests for schools	The contest is for students to create the best poster, essay, or video on a specific water theme.	Water
		Cleanweb Haccelerator	Participants compete to develop clean energy mobile apps and online technologies.	Energy

Notes

1. Bureau of Economic Analysis. 2016. Massachusetts. Available: https://bea.gov/regional/bearfacts/pdf.cfm?fips=25000&areatype=STATE&geotype=3.
2. U.S. Energy Information Administration. 2016. *Massachusetts State energy profile* [Online]. Available: https://www.eia.gov/state/print.php?sid=MA.
3. Executive Office of Energy and Environmental Affairs and Water Resources Commission. 2012. Water conservation standards. Available: http://www.mass.gov/eea/docs/eea/wrc/water-conservation-standards-rev-june-2012.pdf.
4. Union of Concerned Scientists. 2016. Massachusetts's electricity future: Reducing reliance on natural gas through renewable energy. Available: http://www.ucsusa.org/clean-energy/increase-renewable-energy/massachusetts-electricity-future#.WMI0Vzt942w.
5. U.S. Energy Information Administration. 2017. *Massachusetts state profile and energy estimates* [Online]. Available: https://www.eia.gov/state/?sid=MA.
6. Executive Office of Energy and Environmental Affairs. 2017b. *Massachusetts energy profile* [Online]. Available: http://www.mass.gov/eea/energy-utilities-clean-tech/energy-dashboard/mass-energy-profile/.
7. Union of Concerned Scientists. 2016. Massachusetts's electricity future: Reducing reliance on natural gas through renewable energy. Available: http://www.ucsusa.org/clean-energy/increase-renewable-energy/massachusetts-electricity-future#.WMI0Vzt942w.
8. Executive Office of Energy and Environmental Affairs. 2017a. *Agricultural resources facts and statistics* [Online]. Available: http://www.mass.gov/eea/agencies/agr/statistics/.
9. U.S. Energy Information Administration. 2016. *Massachusetts State energy profile* [Online]. Available: https://www.eia.gov/state/print.php?sid=MA.
10. Executive Office of Energy and Environmental Affairs. 2017a. *Agricultural resources facts and statistics* [Online]. Available: http://www.mass.gov/eea/agencies/agr/statistics/.
11. Ibid.
12. Department of Environmental Protection. 2017. Grant announcement. Document title: FFY 17 604b Water quality management planning. Available: http://www.mass.gov/eea/docs/dep/water/ffy17604brfr.pdf.

13. Massachusetts Clean Energy Center. 2017e. *Water innovation* [Online]. Available: http://www.masscec.com/emerging-initiatives/water-innovation-0.
14. Massachusetts Clean Energy Center. 2017a. *Catalyst* [Online]. Available: http://www.masscec.com/innovate-clean-energy/catalyst.
15. Massachusetts Clean Energy Center. 2017c. *InnovateMass* [Online]. Available: http://www.masscec.com/innovate-clean-energy/innovatemass.
16. Massachusetts Clean Energy Center. 2017e. *Water innovation* [Online]. Available: http://www.masscec.com/emerging-initiatives/water-innovation-0.
17. Ibid.
18. Massachusetts Water Resources Authority. 2016. *Poster, writing and video contests for the 2016–2017 school year* [Online]. Available: http://www.mwra.state.ma.us/annual/contest/2017/pre/contestmain.htm.
19. Executive Office of Energy and Environmental Affairs. 2016. MassEVIP application workplace charging program. Available: http://www.mass.gov/eea/docs/dep/air/community/evipwpc-ap.pdf.
20. Department of Energy Resources. 2014. Program opportunity notice. Massachusetts pathways to zero grant program. Available: http://www.mass.gov/eea/docs/doer/procurement/pathways-to-zero-pon.pdf.
21. Massachusetts Clean Energy Center. 2016. Commonwealth wind program manual community and commercial wind projects. Available: http://files.masscec.com/get-clean-energy/govt-np/wind/CommWindProgramManual2016.pdf.
22. DSIRE. 2016. *Commonwealth organics-to-energy program* [Online]. Available: http://programs.dsireusa.org/system/program/detail/5039.
23. Massachusetts Clean Energy Center. 2015. Massachusetts clean energy industry report. Available: http://www.masscec.com/2015-massachusetts-clean-energy-industry-report.
24. Energy and Environmental Affairs. 2015. 2015 update Massachusetts clean energy and climate plan for 2020. Available: http://www.mass.gov/eea/docs/eea/energy/cecp-for-2020.pdf.
25. Ibid.
26. Energy and Environmental Affairs. 2016. *Baker-Polito Administration announces investment in Commonwealth building energy intelligence program* [Online]. Available: http://www.mass.gov/eea/pr-2016/investment-in-commonwealth-building-energy-intelligence.html.

27. Massachusetts Clean Energy Center. 2017b. *Clean energy workforce training capacity building curricula* [Online]. Available: http://www.masscec.com/clean-energy-workforce-training-capacity-building-curricula.

28. Massachusetts Clean Energy Center. 2015. Massachusetts clean energy industry report. Available: http://www.masscec.com/2015-massachusetts-clean-energy-industry-report.

29. Ibid.

30. Massachusetts Clean Energy Center. 2017d. *Learn and earn* [Online]. Available: http://www.masscec.com/work-clean-energy/training-organization/learn-earn.

31. Massachusetts Clean Energy Center. 2015. Massachusetts clean energy industry report. Available: http://www.masscec.com/2015-massachusetts-clean-energy-industry-report.

32. Ibid.

33. Energy and Environmental Affairs. 2017d. *APR program objectives and benefits* [Online]. Available: http://www.mass.gov/eea/agencies/agr/land-use/apr-program-objectives-and-benefits.html.

34. Energy and Environmental Affairs. 2017c. *APR program criteria and considerations* [Online]. Available: http://www.mass.gov/eea/agencies/agr/land-use/apr-application-criteria-and-considerations.html.

35. Massachusetts Department of Agricultural Resources. 2016. APR improvement program. Reinvesting in farmland protected by an Agricultural Preservation Restriction (APR). Available: http://www.mass.gov/eea/docs/agr/programs/aip/mdar-brochure-aip.pdf.

36. Massachusetts Farm Energy Program. 2017b. *Get funding* [Online]. Available: http://massfarmenergy.com/get-funding/.

37. Massachusetts Farm Energy Program. 2017a. *Get an energy audit* [Online]. Available: http://massfarmenergy.com/get-an-energy-audit/.

38. Ibid.

39. Energy and Environmental Affairs. 2017b. *Agricultural energy grant program (Ag-energy)* [Online]. Available: http://www.mass.gov/eea/agencies/agr/about/divisions/ag-energy.html.

40. Energy and Environmental Affairs. 2017a. *Agricultural energy special project grant program (ENER-SP)* [Online]. Available: http://www.mass.gov/eea/agencies/agr/about/divisions/ener-special-projects.html.

41. Energy and Environmental Affairs. 2017e. *Farm viability enhancement program (FVEP)* [Online]. Available: http://www.mass.gov/eea/agencies/agr/about/divisions/fvep.html.

42. Department of Agricultural Resources. 2017c. Request for response—AGR UrbanAg-FY17-3 Massachusetts. Available: http://www.mass.gov/eea/docs/agr/urban/mdar-fy2017-urban-agriculture-rfr.pdf.
43. Department of Agricultural Resources. 2017a. Agricultural environmental enhancement program. Available: http://www.mass.gov/eea/docs/agr/programs/aeep/aeep-drought.pdf.
44. Department of Agricultural Resources. 2017b. Massachusetts Food Ventures Program Fiscal Year 2017. Available: http://www.mass.gov/eea/docs/agr/urban/mdar-fy2017-food-ventures-program-rfr.pdf.
45. Massachusetts Farm Energy Program. 2014. Best management practices for renewable energy. Available: http://massfarmenergy.com/wp-content/uploads/2014/03/Renewable.pdf.
46. Commonwealth Quality Massachusetts. 2017a. *The CQP brand* [Online]. Available: http://thecqp.com/main_cpbrand.html.
47. Commonwealth Quality Massachusetts. 2017b. *Program requirements* [Online]. Available: http://thecqp.com/main_requirements.html.

References

Bureau of Economic Analysis. 2016. Massachusetts. https://bea.gov/regional/bearfacts/pdf.cfm?fips=25000&areatype=STATE&geotype=3
Commonwealth Quality Massachusetts. 2017a. *The CQP brand* [Online]. http://thecqp.com/main_cpbrand.html
———. 2017b. *Program requirements* [Online]. http://thecqp.com/main_requirements.html
Department of Agricultural Resources. 2017a. Agricultural environmental enhancement program. http://www.mass.gov/eea/docs/agr/programs/aeep/aeep-drought.pdf
———. 2017b. Massachusetts Food Ventures Program fiscal year 2017. http://www.mass.gov/eea/docs/agr/urban/mdar-fy2017-food-ventures-program-rfr.pdf
———. 2017c. Request for response—AGR UrbanAg-FY17-3 Massachusetts. http://www.mass.gov/eea/docs/agr/urban/mdar-fy2017-urban-agriculture-rfr.pdf
Department of Energy Resources. 2014. Program opportunity notice. Massachusetts pathways to zero grant program. http://www.mass.gov/eea/docs/doer/procurement/pathways-to-zero-pon.pdf

Department of Environmental Protection. 2017. Grant announcement. Document Title: FFY 17 604b Water Quality Management Planning. http:// www.mass.gov/eea/docs/dep/water/ffy17604brfr.pdf

DSIRE. 2016. *Commonwealth organics-to-energy program* [Online]. http://programs.dsireusa.org/system/program/detail/5039

Energy and Environmental Affairs. 2015. 2015 update Massachusetts clean energy and climate plan for 2020. http://www.mass.gov/eea/docs/eea/energy/cecp-for-2020.pdf

———. 2016. *Baker-Polito Administration announces investment in Commonwealth building energy intelligence program* [Online]. http://www. mass.gov/eea/pr-2016/investment-in-commonwealth-building-energy-intelligence.html

———. 2017a. *Agricultural energy special project grant program (ENER-SP)* [Online]. http://www.mass.gov/eea/agencies/agr/about/divisions/ener-special-projects.html

———. 2017b. *Agricultural energy grant program (Ag-Energy)* [Online]. http:// www.mass.gov/eea/agencies/agr/about/divisions/ag-energy.html

———. 2017c. *APR program criteria and considerations* [Online]. http://www. mass.gov/eea/agencies/agr/land-use/apr-application-criteria-and-considerations.html

———. 2017d. *APR program objectives and benefits* [Online]. http://www. mass.gov/eea/agencies/agr/land-use/apr-program-objectives-and-benefits. html

———. 2017e. *Farm viability enhancement program (FVEP)* [Online]. http:// www.mass.gov/eea/agencies/agr/about/divisions/fvep.html

Executive Office of Energy and Environmental Affairs. 2016. MassEVIP application workplace charging program. http://www.mass.gov/eea/docs/dep/air/community/evipwpc-ap.pdf

———. 2017a. *Agricultural resources facts and statistics* [Online]. http://www. mass.gov/eea/agencies/agr/statistics/

———. 2017b. *Massachusetts energy profile* [Online]. http://www.mass.gov/eea/ energy-utilities-clean-tech/energy-dashboard/mass-energy-profile/

Executive Office of Energy and Environmental Affairs and Water Resources Commission. 2012. Water conservation standards. http://www.mass.gov/ eea/docs/eea/wrc/water-conservation-standards-rev-june-2012.pdf

Massachusetts Clean Energy Center. 2015. Massachusetts clean energy industry report. http://www.masscec.com/2015-massachusetts-clean-energy-industry-report

————. 2016. Commonwealth wind program manual community and commercial wind projects. http://files.masscec.com/get-clean-energy/govt-np/wind/CommWindProgramManual2016.pdf

————. 2017a. *Catalyst* [Online]. http://www.masscec.com/innovate-clean-energy/catalyst

————. 2017b. *Clean energy workforce training capacity building curricula* [Online]. http://www.masscec.com/clean-energy-workforce-training-capacity-building-curricula

————. 2017c. *InnovateMass* [Online]. http://www.masscec.com/innovate-clean-energy/innovatemass

————. 2017d. *Learn and earn* [Online]. http://www.masscec.com/work-clean-energy/training-organization/learn-earn

————. 2017e. *Water innovation* [Online]. http://www.masscec.com/emerging-initiatives/water-innovation-0

Massachusetts Department of Agricultural Resources. 2016. APR improvement program. Reinvesting in farmland protected by an Agricultural Preservation Restriction (APR). http://www.mass.gov/eea/docs/agr/programs/aip/mdar-brochure-aip.pdf

Massachusetts Farm Energy Program. 2014. Best management practices for renewable energy. http://massfarmenergy.com/wp-content/uploads/2014/03/Renewable.pdf

————. 2017a. *Get an energy audit* [Online]. http://massfarmenergy.com/get-an-energy-audit/

————. 2017b. *Get funding* [Online]. http://massfarmenergy.com/get-funding/

Massachusetts Water Resources Authority. 2016. *Poster, writing and video contests for the 2016–2017 school year* [Online]. http://www.mwra.state.ma.us/annual/contest/2017/pre/contestmain.htm

U.S. Energy Information Administration. 2016. *Massachusetts state energy profile* [Online]. https://www.eia.gov/state/print.php?sid=MA

————. 2017. *Massachusetts state profile and energy estimates* [Online]. https://www.eia.gov/state/?sid=MA

Union of Concerned Scientists. 2016. Massachusetts's electricity future: Reducing reliance on natural gas through renewable energy. http://www.ucsusa.org/clean-energy/increase-renewable-energy/massachusetts-electricity-future#.WMI0Vzt942w

7

The Green Economy and the Water-Energy-Food Nexus in Ontario

Introduction

Ontario is Canada's second-largest province in area after Quebec, while its population is the largest in Canada at nearly 14 million with 94 percent concentrated in Southern Ontario close to the St Lawrence River and near the US border. The province's long-term, average, annual real GDP growth of 2.1 percent between 2016 and 2040 is slower than the 2.6 percent average growth from 1982 to 2015 as a result of slower projected labour force growth. Ontario's economy over time has evolved with the share of service industry employment as a percentage of total employment increasing from 73.6 percent in 1996 to almost 80 percent in 2016.[1] Ontario's population is projected to grow by 30.1 percent, or almost 4.2 million, over the next 26 years, from 13.8 million in 2015 to 17.9 million by 2041 with the Greater Toronto Area projected to be the fastest growing region of the province with its population increasing by over 2.8 million, or 42.8 percent, to reach almost 9.5 million by 2041.[2]

© The Author(s) 2018
R.C. Brears, *The Green Economy and the Water-Energy-Food Nexus*,
DOI 10.1057/978-1-137-58365-9_7

7.1   Water-Energy-Food Nexus Pressures

Ontario is experiencing a variety of water, energy, and food nexus pressures that are detrimental to the development of a green economy as described below through a variety of examples.

Water

Ontario's water resources include a quarter-million inland lakes, rivers, groundwater resources, and four of the Great Lakes of North America. The Great Lakes provide drinking water for 10 million Canadians and supply the water needs of industries that produce more than half of the country's total manufacturing output.[3] Climate change is likely to pose a threat to both water quantity and quality. As average summer temperatures rise in Ontario, the demand for water for agricultural irrigation as well as household water use will increase. At the same time, shorter winters and altered snowfall patterns will change groundwater quantities and surface runoff patterns as well as lead to warmer lakes and streams.[4]

Energy

In Ontario, over 61 percent of electricity generated is from nuclear power, while hydropower, gas, and wind contribute 20.9 percent, 12.3 percent, and 4.1 percent respectively, with the remainder from biofuel and solar.[5] While electricity consumption province wide has remained flat with no significant growth since 2009, there are some population growth areas where energy needs are growing, particularly in the Greater Toronto Area as well as around Ottawa. For instance, almost 80 percent of Toronto's grid stations will be at capacity by 2019 due to increased population growth and higher electricity demand. This growth in urban energy demand will put additional pressure on the system during periods of peak demand, such as on hot days when air conditioning is turned up high. For example, in Toronto, demand for electricity during these peak periods is expected to grow over 1 percent a year.[6]

Food

Ontario is one of North America's largest and most significant food and beverage processing locations with nearly 3000 registered food and beverage businesses. Ontario is also home to a wide variety of raw ingredients with over 50,000 conventional and organic farms growing over 200 Ontario-produced agricultural commodities with processors purchasing about 65 percent of Ontario's food-related farm production.[7] However, there is evidence that Ontario does not produce enough food to feed itself, with food deficits requiring import of potatoes, apples, beef, chicken, and lamb, among others. In addition, the province imports foods that the province produces itself, for example, Florida-grown vegetables over locally, home-grown vegetables. If these trends were reversed, $250 million would be created for the economy, producing nearly 3500 jobs.[8]

7.2   General Fiscal Tools to Reduce Water-Energy-Food Nexus Pressure

Ontario has implemented a variety of fiscal tools that create interdependencies and synergies between the nexus systems while reducing trade-offs between the systems in the development of a green economy.

Innovation Demonstration Fund

Ontario's Innovation Demonstration Fund is a discretionary, non-entitlement funding programme administered by the Ministry of Economic Development and Innovation that focuses on emerging technologies with a preference towards environmental, alternative energy, bio-products, hydrogen, and other globally significant technologies with each applicant able to apply for funding up to 50 percent but not to exceed $4 million. Overall, the Fund supports pilot demonstrations that will lead to the commercialisation of processes and/or products in Ontario that are globally competitive and innovative green technologies.[9]

7.3   General Non-fiscal Tools to Reduce Water-Energy-Food Nexus Pressures

Ontario has implemented a variety of non-fiscal tools that create interdependencies and synergies between the nexus systems while reducing trade-offs between the systems in the development of a green economy.

Supporting the Sustainability CoLab

Ontario's Green Investment Fund will invest $1 million into Kitchener's Sustainability CoLab Network, which is an organisation that helps small and medium-sized businesses set and achieve targets to reduce their environmental impact while enhancing their competitive advantage and stimulating the development of a low-carbon economy. The funding will be used by Sustainability CoLab to fund local environmental organisations in their network to help businesses undertake initiatives including introducing energy efficiency measures and completing energy retrofits and building audits.

Climate Change Training for Indigenous Communities

The Green Investment Fund will provide $5 million to the Ontario Centre for Climate Impacts and Adaptation Resources, in partnership with the Ontario First Nations Technical Services Corporation. The funding will help Indigenous communities develop climate change adaptation plans, build technical capacity to take advantage of economic opportunities arising from a low-carbon economy, and create a Northern Ontario climate change impact study using data from the adaptation plans. The fund will also commit $8 million towards the development of micro-grid solutions in First Nations communities. This will support economic growth by reducing reliance on diesel fuel and enable stable and predictable sources of power.[10]

7.4 Water: Fiscal Tools

Ontario has implemented a variety of water-related fiscal tools that create interdependencies and synergies between the nexus systems while reducing trade-offs between the systems in the development of a green economy.

Ontario Water Resources Act Permits

Ontario requires a permit if a water user plans to take 50,000 or more litres of water per day from the environment including lakes, streams, rivers, ponds, or groundwater with a permit's current duration lasting for ten years. The permit will not be issued for activities that would negatively affect existing users, negatively affect the environment, or remove water from a watershed that already has a high level of use. Nonetheless, water users are exempt from requiring a permit if they take less than 50,000 litres or take water for: livestock, poultry, irrigation, and frost protection for agriculture; domestic purposes (e.g. home gardens and lawns); firefighting or other emergency purposes; wetland conservation or a weir constructed prior to March 2016; passive and/or active in-stream diversions for construction purposes; or receive water supplied by another water user with a valid permit to take water. For a permit to be issued, the following factors need to be considered:

- *Natural ecosystem functions*: For example, the potential impact of the taking on minimum stream flows, natural water level/flow variations, habitat that depends on levels/flows and interrelationships between groundwater and surface water
- *Water availability*: For example, the impact on water balance and sustainable aquifer yield, existing water uses, low water conditions, whether the taking is in a high-use or medium-use watershed, any planned municipal use approved
- *Issues related to the use of water*: For example, whether water conservation is implemented in accordance with best water management standards for the relevant sector, the purpose of water use

Water Taking Classification

The types of permits applied for depend on their anticipated risk to the environment, with three types of permits available:

- Category 1: Low risk and includes renewals where this is no history of complaints, with an application fee of $750
- Category 2: For water takings with a greater potential to cause adverse environmental impact (requiring a scientific evaluation/ study to be completed by a qualified person), with an application fee of $750
- Category 3: Considered high risk (requiring a scientific evaluation/ study to be completed by a qualified person), with an application fee of $3000[11]

7.5   Water: Non-fiscal Tools

Ontario has implemented a variety of water-related non-fiscal tools that create interdependencies and synergies between the nexus systems while reducing trade-offs between the systems in the development of a green economy.

Water Opportunities Act: Water Technology Acceleration Project

Ontario's Water Opportunities Act 2010 aims to foster innovative water, wastewater, and stormwater technologies, services, and practices in both the private and public sectors, which will in turn create opportunities for clean-technology jobs in Ontario and conserve and sustain water resources for present and future generations. As part of the Act, the Water Technology Acceleration Project was formed with the objectives of:

- Assisting in promoting the development of Ontario's water and wastewater sectors
- Assisting Ontario's water and wastewater sectors in increasing their capacity to:
  - Develop, test, demonstrate, and commercialise innovative technologies and services for the management of water and wastewater
  - Expand their business opportunities nationally and internationally
- Providing a forum for governments, the private sector, and academic institutions to exchange information and ideas on how to make Ontario a leading jurisdiction in the development and commercialisation of innovative technologies and services for the treatment and management of water and wastewater
- Encouraging collaboration and cooperation in Ontario's water and wastewater sectors
- Assisting, if requested by the Minister, in the development, certification, labelling and verification programmes for water and wastewater technologies and services
- Providing the Minister with advice on what actions the Government of Ontario should take in fostering the development of Ontario's water and wastewater sectors[12]

Legislating Conservation Authorities

Conservation Authorities, legislated by the Conservation Authorities Act (1946), are non-profit organisations that are mandated to ensure the conservation, restoration, and responsible management of Ontario's water, land, and natural habitats through programmes that balance human, environmental, and economic needs. The objectives of the Conservation Authorities are to:

- Ensure that Ontario's rivers, lakes, and streams are properly safeguarded, managed, and restored
- Protect, manage, and restore Ontario's woodland, wetlands, and natural habitat

- Develop and maintain programmes that will protect life and property from natural hazards including flooding and erosion
- Provide opportunities for the public to enjoy, learn from, and respect Ontario's natural environment

Conservation Authority Governance Arrangements

Each Conservation Authority has their own Board of Directors that are comprised of members appointed by local municipalities with the majority of Board representatives elected municipal officials. In total 31 Conservation Authorities operate in southern Ontario while an additional five Conservation Authorities deliver services in northern Ontario. Around 90 percent of Ontario's more than 13 million residents live in a watershed managed by a Conservation Authority. Funding for Conservation Authorities is derived from a variety of sources: on average 48 percent comes from municipal levies, 40 percent comes from self-generated revenues, 10 percent comes from provincial grants and special contracts, and 2 percent from federal grants or contracts.

Water Source Protection Areas

Ontario has established source protection areas and created a multi-stakeholder source protection committee for each area. The committee identifies significant existing and future risks to their drinking water sources and develops plans to mitigate these risks. Ontario has funded the entire source protection planning process, spending over $250 million to invest in technical and scientific studies, develop local plans, and encourage voluntary action by landowners.[13]

The Ontario Drinking Water Stewardship Program

The Ontario Drinking Water Stewardship Program provided $24.5 million to assist over 3000 local actions by landowners to protect water supplies including:

- Implementing runoff and erosion control measures
- Inspecting and upgrading septic tanks
- Closing or upgrading wells[14]

Source Water Protection Map

In 2016, Ontario launched the Source Water Protection Map, which is an interactive tool that provides the first province-wide view of more than 970 wellhead protection areas and 150 intake protection zones within the source protection areas (the places where drinking water comes from). The public can access over 20 layers of information and do customised searches. Another tool is the installation of road signs on provincial and municipal roads to inform the public and early responders of the location of vulnerable drinking water supplies. Ontario also uses social media to reach new audiences with an interest in protecting water sources and facilitates online conversations about source protection.[15]

7.6   Energy: Fiscal Tools

Ontario has implemented a variety of energy-related fiscal tools that create interdependencies and synergies between the nexus systems while reducing trade-offs between the systems in the development of a green economy.

Electric Vehicle Incentive Program

In 2010, Ontario introduced the Electric Vehicle Incentive Program (EVIP) to promote the adoption of battery electric and plug-in hybrid vehicles. A range of subsidies are available including:

- Vehicles with a battery capacity from 5 to 16 kilowatt-hours (kWh) can receive an incentive between $6000 and $10,000 based on the vehicle's battery size

- Vehicles with batteries larger than 16 kWh receive an additional $3000 incentive
- Vehicles with five or more seats are eligible for an additional $1000 incentive[16]

Electric Vehicle Charging Incentive Programme

Ontario is supporting the uptake of EVs by offering up to $1000 to help homes and businesses purchase and install Level 2 charging stations. To be eligible, applicants must of received the provincial rebate for the purchase of an EV. The rebate covers 50 percent of the purchase cost, up to $500, and 50 percent of the installation cost, up to $500.[17]

Electric Vehicle Chargers Ontario Public–Private Partnership Funding

Ontario's Electric Vehicle Chargers Ontario grant programme works with public and private sector partners to create a network of fast-charging EV stations in cities, along highways and at workplaces, condominiums, and public places across Ontario. In total 27 partners are involved in installing the chargers at a wide variety of locations including McDonald's, Tim Horton's restaurants, IKEA parking lots, Pearson International Airport, and at numerous municipal properties and businesses across Ontario. Over the coming years about 300 Level 2 and 200 Level 3 charging stations will be installed across Ontario at over 250 locations. By February 2016, over 200 applications were received from businesses, municipalities, Aboriginal communities or organisations, local distribution companies, non-governmental organisations, conservation authorities, and other entities. Applications were evaluated on optimal connectivity, site or corridor selection, innovativeness, cost efficiency, and effectiveness and reasonableness.[18]

Save on Energy: Audit Funding

Ontario's Save on Energy provides an audit funding incentive for businesses to cover up to 50 percent of the cost of an energy audit. The energy audit programme is designed to be the first step in understanding how energy is being used in a building.

Audit Funding for Businesses

All types of business customers, including commercial, institutional, industrial, agricultural, and multi-residential facilities including social housing, are eligible for incentives (Table 7.1) to complete energy audits addressing the potential for energy savings through equipment replacement, operational practices, or participation in building systems and envelope projects. There are two parts to the audit process: First, businesses

Table 7.1 Save on Energy audit funding for businesses

Audit type	For building owners	
	Buildings up to 30,000 square feet	Buildings larger than 30,000 square feet
Electrical survey and analysis	$0.10/square feet up to a maximum of 50%	$3000 for the first 30,000/square feet and $0.05/square feet up to $25,000 or 50% of audit costs thereafter
Detailed analysis of capital-intensive modifications	Buildings larger than 50,000 square feet $0.05/square feet up to $10,000 or 50% of audit costs	
	For tenants or building owners	
Building systems audit	Up to $5000 or 50% of audit costs	
Electricity survey and analysis	$0.03/square feet up to $7500 or 50% of audit costs	

Save on Energy. 2014. Audit funding. https://saveonenergy.ca/Business/Program-Overviews/Audit-Funding.aspx

have an electrical survey and analysis conducted that provides information on the financial benefits of installing and upgrading a variety of energy-efficient equipment. Second, businesses have a detailed analysis of capital-intensive modifications.

Audits for Tenants or Building Owners

Tenants of facilities and building owners can have a building systems audit that provides an analysis of achievable savings with modifications to balancing and optimising auxiliary fans, pumps, compressors, and domestic water. Instead of the building systems audit tenants can opt for an electricity survey and analysis that enables tenants of facilities to identify potential energy-efficient equipment upgrades and replacements and the financial benefits associated with each project.[19]

Ontario's Net Metering Credit Programme

Ontario's Net Metering programme is available to Ontario customers who generate electricity primarily for their own use and from a renewable source, including bioenergy, wind, solar photovoltaic, and water-power, using equipment of maximum cumulative capacity up to 500 kilowatts (kW) in size. Net metering is a billing arrangement where customers can send electricity generated from renewable sources to the electrical grid for a credit towards their electricity bill. The customer's local distribution company will then subtract the value of the electricity supplied to the grid from the value of what was taken from the grid over a billing period. If a customer supplies more electricity than was taken they will then receive a credit that can be carried over to lower future electricity bills.[20]

Feed-in Tariff for Small Businesses

Ontario supports the development of small-scale renewable energy through the Feed-in Tariff (FIT) programme, which is administered by the Independent Electricity System Operator and through the Net Metering programme administered by local distribution companies. The FIT programme offers participants stable prices under fixed-term contracts for energy generated from renewable sources including bioenergy, wind, solar, and photovoltaic and waterpower. Under the FIT programme, participants are paid a guaranteed price over a 20-year term (40 years for waterpower projects) for the electricity project to produce and deliver to the province's electricity grid.[21]

Smart Grid Funding Grants

Ontario's Smart Grid Fund aims to accelerate the development of the province's smart grid industry by providing targeted financial support for projects that advance the development of the smart grid in Ontario and provide economic opportunities including the creation of new jobs. The Fund's specific objectives are to:

• Develop and advance the smart grid in Ontario by advancing one or more of the smart grid objectives in the focus areas of consumer control, power system flexibility, and adaptive infrastructure
• Create economic development opportunities including jobs for Ontario
• Reduce risk and uncertainty of electricity sector investments by enabling utilities and other electricity industry stakeholders to develop, test, and evaluate smart grid technologies and business models

Smart Grid Fund Grants

Successful applicants with projects that have a maximum time frame of two years will receive funding up to $2 million per project with the funding allowed to cover up to 50 percent of eligible project costs. The Fund also allows projects to receive funding from other federal and Ontario government programmes up to 75 percent of total eligible costs.[22]

7.7   Energy: Non-fiscal Tools

Ontario has implemented a variety of energy-related non-fiscal tools that create interdependencies and synergies between the nexus systems while reducing trade-offs between the systems in the development of a green economy.

Greener Diesel Regulation

Ontario requires fuel companies to provide more environmentally friendly diesel fuels, known as bio-based diesel fuels, to help reduce air pollutants, improve air quality, and reduce GHG emissions in the transportation sector. There are two types of bio-based diesel fuels: biodiesel which is a renewable fuel made from vegetable oils, recycled frying oils, and animal fats; and renewable diesel which is made from the same materials as biodiesel but it is processed differently. Over the period 2014–2017, Ontario is setting minimums for the amount of bio-based diesel distributed, used, and/or sold in Ontario:

- In 2014/2015, 2 percent of the total volume of diesel fuel must be bio-based with the bio-based component of the blend having 30 percent lower GHG emissions than standard petroleum diesel.
- In 2016, 3 percent of the total volume of diesel fuel must be bio-based with the bio-based diesel component of the blend having 50 percent lower GHG emissions than standard petroleum diesel.
- In 2017, 4 percent of the total volume of diesel fuel must be bio-based with the bio-based diesel component of the blend having 70 percent lower GHG emissions than standard petroleum diesel.[23]

Green Licence Plate Program

Green licence plates have ongoing access to high-occupancy vehicle lanes and no-cost access to high-occupancy toll lands, even if there is only one person in the car. Individuals, businesses, and fleets are eligible for green licence plates if they are using vehicles that meet the following criteria:

- Plug-in hybrid electric vehicles and battery electric vehicles are eligible for the EVIP
- Used 2010 or later model year plug-in hybrids and battery electric vehicles
- Plug-in hybrids, battery electric vehicles, and hydrogen fuel cell vehicles currently operating in Ontario in limited numbers as part of pilot studies or test programmes[24]

Energy and Water Reporting and Benchmarking

Ontario's Ministry of Energy's Energy and Water Reporting and Benchmarking (EWRB) requires that all privately owned buildings that are 50,000 square feet and larger, including commercial, multi-residential with more than ten residential units and some industrial buildings/properties, report their energy, water, GHG emissions as well as other building characteristic information to the Ministry on an annual basis with the Ministry disclosing some of this information to the public and developing reports summarising key findings. Building owners are required to report their data to the Ministry of Energy by July of every year. The reporting requirement will be phased in over three years, beginning 1 July 2018, with buildings that are greater than or equal to 250,000 square feet (summarised in Table 7.2).

Table 7.2 Energy and Water Reporting and Benchmarking

Phased mandatory reporting	Commercial and industrial	Multi-unit residential
By July 1, 2018	250,000 square feet and larger	Not required to report in the first year
By July 1, 2019	100,000 square feet and larger	100,000 square feet and larger
By July 1, 2020	50,000 square feet and larger	50,000 square feet and larger

Ministry of Energy. 2017a. *Energy and water reporting and benchmarking for large buildings* [Online]. http://www.energy.gov.on.ca/en/ontarios-ewrb/

A Public Database to Compare Buildings

The Ministry of Energy will annually disclose a subset of the data reported by the building owners on Ontario's Open Data website. The data will include property identification and building performance information such as ENERGY STAR score; energy, water, and GHG intensity; building age; and confirmation of whether the data was verified by an accredited/certified professional. This will encourage building owners to compete with one another and strive to improve their building performance each year. It also will enable property and financial markets to compare building performance and value efficient buildings, enabling the market to drive further energy- and water-efficiency investments. To let building owners gain experience with reporting their data, the Ministry will not publicly disclose the first year of reported data for each of the three roll-out phases. Overall, the initiative is summarised in Table 7.3

Table 7.3 Energy and water reporting and benchmarking policy elements

Policy element	Description
Building types	Commercial, multi-residential units 50,000 square feet and above to be included. Most industrial buildings, that is, manufacturing facilities and all agricultural facilities, would not be included.
Annual reporting	ENERGY STAR Portfolio Manager will be used to report building data on an annual basis, including monthly energy and water consumption (and performance data where available), GHG emissions, building characteristic information, for example, gross floor area. Electricity, natural gas, and water utilities will also be required to make whole building, aggregated, consumption data available to building owners so they can comply with the requirement.

(continued)

Table 7.3 (continued)

Policy element	Description
Annual disclosure	Public disclosure on Ontario's Open Data website one year after initial reporting for each of the three phases. Some data including site/source energy use, total GHG emissions, and gross floor area on a building by building basis will not be disclosed publicly.
Data verification	Building owners are required to confirm in Portfolio Manager that the reported data is accurate. Verification by a third party will not be required under the proposed programme.
Reports	The Ministry of Energy will publish reports summarising key findings.
Conservation and demand management plans	At this stage there is no requirement for conservation and demand management plans in the initial years of reporting; however, the Ministry will consider requiring these plans on a targeted basis in future years based on results.

Government of Ontario. 2016. *Large building energy and water reporting and benchmarking* [Online]. https://www.ebr.gov.on.ca/ERS-WEB-External/displaynoticecontent.do?noticeId=MTI3ODY0&statusId=MTkzMTc3

Ontario's Green Button Initiative

The Green Button Initiative, launched in 2014, empowers consumers by providing easier access to their own electricity data and allowing them to securely share their data with mobile and web-based apps. The apps can help Ontario homeowners and businesses better understand and manage their electricity use so they can make informed decisions and take action to reduce their energy bills. Residential and commercial customers have a choice of four apps to identify their electricity data and better manage their usage. In 2015, a new collection of electricity management apps for Ontarians were made available as part of the second phase of the Ontario Green Button Connect My Data pilot. The apps are summarised in Table 7.4

Table 7.4 Green button smart phone apps

Year	App	Description
2014	MyEyedro by Eyedro	A fun, fast, and free way to manage electricity use. At home or on the go, users can see how their electricity usage compares against family and friends and find out how much they can save when they take action.
	GOODcoins by Zerofootprint	Rewards customers for tracking their electricity and achieving electricity reduction targets. Customers earn rewards for using less electricity.
	BuiltSpace	Provides a comprehensive building asset, energy, and service management solution to optimise energy and operational efficiency across diverse building portfolios.
	Energent	Incorporates statistical energy modelling capabilities to track, monitor, and assess customers' energy data and facility energy performance.
2015	HomeBeat by Bidgely	An energy advisor web and mobile app that provides residential customers with advice to help them understand energy use by showing the impact of individual appliances on the bill and providing them with tips to reduce consumption.
	Presence Pro Energy by People Power	Delivers real-time, whole-home energy monitoring and smart plug control giving residential and small business users the ability to manage their electric use from a smartphone or tablet.
	Wattsly	A personalised energy 'butler' that helps people monitor and lower their electricity usage, and their energy bills.
	Stream by Energy Profiles Limited	A comprehensive utility data management system that enables property owners and managers to understand their utility use and cost at both the portfolio and building levels.

Ministry of Energy. 2017b. *Ontario's green button initiative* [Online]. http://www.energy.gov.on.ca/en/ontarios-electricity-system/green-button/

7.8 Food: Fiscal Tools

Ontario has implemented a variety of food-related fiscal tools that create interdependencies and synergies between the nexus systems while reducing trade-offs between the systems in the development of a green economy.

Food Donation Tax Credit

Ontario has introduced a Food Donation Tax Credit in which farmers who give agricultural products to eligible community food programmes receive a tax credit of 25 percent for the fair market value of the agricultural products they donate. Agricultural products eligible for the tax credit include fruits, vegetables, meat, eggs or dairy products, fish, grains, pulses, herbs, honey, maple syrup, mushrooms, nuts, or anything else that is grown, raised, or harvested on a farm that may be legally sold, distributed, or offered for sale in Ontario at a place other than the premises of its producer.[25]

Greenbelt Fund

The Greenbelt Fund, with funding from the Province of Ontario, is investing $6 million to increase local food procurement within the broader public sector, increase market access for small- and medium-sized firms and processors, and improve understanding of where local food can be found, what local foods are available, and how to use them. This will be achieved through the Greenbelt Fund's three main grant categories, with each successful grant application receiving funding up to 50 percent of the project's eligible costs:

Local Food Literacy Grant Stream

This stream aims to increase awareness and knowledge of local food, its availability, and local food skills to maximise local food usage among consumers, retailers, wholesalers, foodservice operators, associations, and governments as well as increase farmers' and commodity organisations' ability to communicate the value of their products to retailers, wholesalers, foodservice operators, and consumers.[26]

Broader Public Sector Grant Stream

This stream focuses on increasing the amount of Ontario food purchased by public institutions specifically municipal, colleges, universities, school boards, and hospital foodservices, as well as enhancing the capacity of the agri-food sector to access the broader public foodservice industry to highlight the availability and increase the purchase of local products.[27]

Market Access Food Grant Stream

This stream focuses on increasing market access for farmers and producers in order to increase local food choices. This is achieved by:

- Developing food hubs and regional aggregators to increase the number of farmers that have access to retail, wholesale, foodservice, and farmers' markets.
- Encouraging value-added local food processing.
- Developing new and emerging markets by assisting with business intelligence, networking, and relationship building. This includes the development of new tools on Ontario *fresh*.ca—the Greenbelt Fund's online community and database that connects over 2500 local food buyers, processors, and sellers across the province.[28]

Fresh from the Farm

Fresh from the Farm is a partnership between the Ontario Ministry of Education, Ontario Ministry of Agriculture, Food and Rural Affairs (OMAFRA), Dietitians of Canada, and the Ontario Fruit and Vegetable Growers' Association that provides schools with the opportunity to raise funds by selling fresh, Ontario-grown fruit and vegetables to the community. The programme's financial model is:

- 40 percent of sales go directly to the school (a $75 delivery fee is charged to cover distribution costs)

- 50 percent of sales will be returned to Ontario's farmers
- 10 percent is retained by the Fresh from the Farm for programme sustainability

Growing Forward 2

Growing Forward 2 is a project-based cost-share funding assistance programme that supports agriculture, agri-food, and agri-based bio-products in Ontario. The programme, administered by the Ontario Soil and Crop Improvement Association on behalf of the OMAFRA, provides support in six focus areas of:

1. Environment and climate change adaptation
2. Animal and plant health
3. Market development
4. Labour productivity enhancements
5. Assurant systems (food safety, traceability, animal welfare, and weather risk mitigation)
6. Business and leadership development

Growing Forward 2 for Organisations and Collaborations

Growing Forward 2 supports training and benchmarking as well as improvements in equipment and technology that improve nutrient use efficiency, nutrient application, resource management, and management of water. Organisations and collaborations can also undertake projects to reduce waste and manage inputs including water and energy more efficiently to improve their overall operational, economic, and environmental performance. The fund provides assistance up to 50 percent cost-share funding of total eligible costs. Funding requests over 50 percent must demonstrate how the project is innovative and may receive up to 75 percent of total eligible costs.[29]

Growing Forward 2 for Processors

Growing Forward 2 offers project-based cost-sharing funding assistance to processors who are actively processing, modifying, or transforming agricultural commodities, food, beverages, or agricultural-based bio-products in Ontario. The maximum cost-share funding that a single processor business can receive over a five-year time frame is $350,000. In the area of environment and climate change adaptation the programme provides funding for projects that focus on improving the use of energy and water resources and reducing wastewater and solid waste output. The maximum available cost-share per project category is:

- Assessment or audit, planning, skills development, and training, $5000
- Measure and benchmark resource and waste management efficiency, $20,000
- Equipment or technology, $80,000
- Waste stream equipment or technology upgrades, $100,000[30]

Growing Forward 2 for Producers

The Growing Forward 2 programme provides funding for weather risk mitigation projects under the environment and climate change adaptation focus area. These projects will focus on on-farm practices and strategies to address cold weather risk for perennial, edible horticulture crop producers except for producers of wine grapes. Producers are encouraged to take advantage of free workshops and training opportunities that include a two-day workshop on environmental farm management. The maximum cost-share funding that a single producer can receive over a five-year time frame is $350,000. The province may provide:

- 50 percent cost-share funding up to a pre-determined cost-share cap for projects under the assessment of audit, planning, skills development, and training category

- Up to 35 percent, up to a pre-determined cost-share cap, for remaining project categories
- Enhance cost-share for projects that demonstrate high levels of innovation[31]

7.9   Food: Non-fiscal Tools

Ontario has implemented a variety of food-related non-fiscal tools that create interdependencies and synergies between the nexus systems while reducing trade-offs between the systems in the development of a green economy.

Increasing Local Food Literacy

Foodland Ontario is the government's flagship programme to help increase local food literacy. For over 40 years the programme has helped consumers identify locally grown food and appreciate the diversity of local products. Since 1977, it has established over 1200 logo placements with producers, retailers, and food service operators. In 2015, Foodland Ontario distributed nearly ten million pieces of point-of-sale material and information resources to retailers, on-farm markets, farmers' markets, and food service providers. In addition, the organisation held 135 events over the province as well as generated over 1000 print articles and made over 200 television appearances.[32]

7.10   Case Study Summary

Ontario has implemented a variety of initiatives to create interdependencies and synergies between the nexus systems while reducing trade-offs between the systems in the development of a green economy.

General

- Ontario has an innovation demonstration fund that supports various environmental and renewable energy technologies that have the potential to be commercialised not just in Ontario but globally.
- The province's Green Investment Fund supports a non-profit organisation that helps SMEs reduce their environmental impacts while enhancing their competitiveness. The fund also supports Indigenous communities implement low-carbon economy initiatives.

Water

- To reduce pressures on water resources, Ontario has a water resources permitting system for uses that may have an adverse impact on the environment with the cost of the permit dependent on the level of risk.
- To increase the growth of the province's water technologies sector and create jobs, the Water Technology Acceleration Project was developed to encourage collaboration between the public and private sectors as well as between private sector actors on water- and wastewater-related technologies.
- To protect water resources, Ontario has given powers to non-profit organisations to ensure the responsible management of water while balancing human and economic needs. The province funds a portion of these organisations' operating budgets with the remainder coming from municipalities as well as self-generated funding sources.
- Ontario has established water source protection areas and to enhance awareness of these areas has established an interactive water source protection map for the public.

Energy

- To enhance the uptake of EVs, Ontario has established an incentive programme that subsidises a range of EVs in addition to an incentive for homes and businesses to install charging stations.

- To rapidly expand the EV charging network Ontario has set up a public–private fund that involves large partners installing fast-charging EV stations across the province.
- To encourage energy savings the province provides funding for energy audits for businesses as well as tenants of facilities.
- Ontario provides a net metering programme where customers who generate excess renewable energy can send it back to the grid and receive a credit on their electricity bill. In addition, the province offers a feed-in tariff for small-scale businesses that produce renewable energy for the province's grid, guaranteeing the business a stable price over a fixed-term period.
- To encourage renewable fuel for transportation, Ontario is setting minimum amounts of biodiesel to be distributed and sold in the province.
- Ontario requires all large privately owned buildings report their energy and water usage to the Ministry of Energy each year. In addition, the Ministry will annually disclose a subset of the data submitted so the property and financial market can compare building performance and value efficient buildings.
- To help homeowners save energy, Ontario's Green Button Initiative enables customers to access their own electricity data and view on their mobile phones with a range of phone apps designed to compare energy usage with others, reward customers for saving energy as well as provide energy-saving tips.

Food

- Ontario has a food donation tax credit to encourage farmers to donate produce to community food programmes.
- To increase local food procurement, the province is providing a variety of grants to raise awareness and knowledge of local food and its availability, enhance the purchasing of local produce by public institutions, and develop food hubs for local farmers to sell their produce.
- To encourage the purchasing of locally produced healthy food, the province enables schools to sell locally grown produce to the community

Table 7.5 Ontario case summary

Tool	Tool type	Policy title	Description	WEF sectors addressed
Fiscal	Market-based instruments and pricing	Ontario Water Resources Permits	Large water users must acquire a permit with the permit fee determined by the level of anticipated risk to the environment.	Water
		Net metering credit programme	Customers can send electricity generated from renewable sources to the electrical grid for a credit towards their electricity bill.	Energy, food
		Feed-in Tariff	Offers participants stable prices under fixed-term contracts for energy generated from renewable sources.	Energy, food
		Fresh from the Farm	Schools can raise money by selling local food with a portion of sales returned to local farmers.	Food
	Environmental taxes	Food Donation Tax Credit	Farmers can receive a tax credit for donating agricultural produce to community food programmes.	Food
	Financial incentives	Innovation Demonstration Fund	The Fund supports pilot demonstrations involving emerging environmental-, alternative energy-, bio-products-, hydrogen-related technologies.	Water, energy, food
		Electric Vehicle Incentive Program	A range of subsidies are available for the purchase of electric vehicles.	Energy

(continued)

Table 7.5 (continued)

Tool	Tool type	Policy title	Description	WEF sectors addressed
		Electric Vehicle Incentive Program	EV owners can receive an incentive for installing a home charging station.	Energy
		Save on Energy audit funding	All types of business customers, as well as tenants of facilities and building owners, are eligible for energy audit funding.	Energy, food
		Smart Grid funding	Grants are available for projects that advance the development of the smart grid in Ontario.	Energy
		Growing Forward 2 for organisations and collaborations	Funding assistance for training and benchmarking in improving agricultural nutrient management.	Food, water, energy
		Growing Forward 2 for processors	Funding for food processors to improve energy and water usage.	Food, water, energy
		Growing Forward 2 for producers	Funding for agricultural climate/weather risk mitigation projects.	Food, water
		Local food literacy grant	Aims to increase awareness and knowledge of local food, its availability, and local food skills to maximise local food usage.	Food
		Broader public sector local food grant	Aims to increase the amount of local food purchased by public institutions.	Food

(continued)

Table 7.5 (continued)

Tool	Tool type	Policy title	Description	WEF sectors addressed
Non-fiscal	Regulations	Market food access grant	Increases market access for farmers and producers.	Food
		Energy and Water Reporting and Benchmarking	All privately owned buildings need to report their energy and water usage annually.	Energy, water
		Greener diesel regulation	Ontario is setting minimum criteria for the amount of bio-based diesel distributed, used, and/or sold in Ontario.	Energy, food
		Green Licence Plate Program	Electric vehicles and hybrids can use high-occupancy vehicle lanes.	Energy
	Public education and skills development	Climate change training for Indigenous communities	The Green Investment Fund is supporting climate change education that helps Indigenous communities develop adaptation plans and develop the low-carbon economy.	Water, energy, food
	Public-private partnerships	Legislating Conservation Authorities	Non-profit organisations are mandated by law to help protect waterways.	Water
		Ontario Drinking Water Stewardship Program	Financial assistance is provided to assist landowners to protect drinking water supplies.	Water
		Electric Vehicle Chargers Ontario	Public–private partnerships are developed to create electric vehicle–charging station networks.	Energy

(continued)

Table 7.5 (continued)

Tool	Tool type	Policy title	Description	WEF sectors addressed
	Stakeholder participation	Supporting the Sustainability CoLab	The Green Investment Fund is supporting an organisation that helps SMEs reduce environmental impacts and stimulate the low-carbon economy.	Water, energy, food
		Water source protection committees	Committees identify existing and future risks to their drinking water sources and develop plans to mitigate these risks.	Water
	Information and awareness-raising	Green Button Initiative	Consumers can access their own electricity data to make informed decisions on how to reduce their consumption.	Energy
	Clustering policies	Water Technology Acceleration Project	The Project aims to assist in developing the water sector, facilitate the testing of new technologies, and encourage the sharing of knowledge.	Water
	Resource mapping	Source Water Protection Map	The interactive tool provides a province-wide view of source protection areas and water intake areas.	Water
	Public procurement	Greenbelt Fund	Ontario is increasing local food procurement within the public sector through a public sector grant.	Food

with half of the sales returned to Ontario's farmers and the remainder going directly to the schools with a small portion used to maintain the programme.

- Ontario's Growing Forward 2 programme is a cost-sharing assistance programme that enhances agricultural productivity across the province, with funds available to organisations and collaborations that enhance water and energy efficiency and nutrient use efficiency as well as funding for processors to enhance water and energy efficiency as well as funds for producers to mitigate climate risks.
- To increase local food literacy, Ontario provides point-of-sale materials and other information resources to retailers and food service providers.

Overall, Ontario uses a variety of fiscal and non-fiscal tools to create interdependencies and synergies between the nexus systems while reducing trade-offs between the systems in the development of a green economy. These tools are summarised in Table 7.5.

Notes

1. Ontario Ministry of Finance. 2017a. Economic trends and projections. *Ontario's long-term report on the economy* [Online]. http://www.fin.gov. on.ca/en/economy/ltr/.
2. Ontario Ministry of Finance. 2017b. Ontario Population Projections Update, 2015–2041.
3. WaterTAP. 2017. *The story* [Online]. http://watertapontario.com/ the-story/.
4. Ministry of the Environment and Climate Change. 2016. *Minister's Annual Report on Drinking Water 2016* [Online]. https://www. ontario.ca/page/ministry-environment-and-climate-change-ministers-annual-report-drinking-water-2016.
5. Ontario Energy Board Commission. 2016. Ontario energy report Q3 2016. http://www.ontarioenergyreport.ca/pdfs/5993_IESO_Q3OER2016_Electricity_EN.pdf.
6. Mowat Centre. 2013. Background report on the Ontario energy sector.
7. Invest Ontario. 2017. *Food and beverage manufacturing* [Online]. http:// www.investinontario.com/food#diverse.

8. Kubursi, A., Cummings, H., Macrae, R. & Kanaroglou, P. 2015. Dollars & sense: Opportunities to strengthen southern Ontario's food system. http://www.greenbelt.ca/dollars_and_sense_opportunities_2015.

9. Mentorworks. 2017. *Ontario innovation demonstration fund* [Online]. https://www.mentorworks.ca/what-we-offer/government-funding/capital-investment/idf/.

10. Government of Ontario. 2017a. *Green investment fund* [Online]. https://www.ontario.ca/page/green-investment-fund.

11. Government of Ontario. 2017c. *Permits to take water* [Online]. https://www.ontario.ca/page/permits-take-water.

12. Government of Ontario. 2010. *Water Opportunities Act, 2010, S.O. 2010, c. 19, Sched. 1* [Online].

13. Government of Ontario. 2017d. *Source protection* [Online]. https://www.ontario.ca/page/source-protection.

14. Ibid.

15. Ministry of the Environment and Climate Change. 2016. *Minister's Annual Report on Drinking Water 2016* [Online]. https://www.ontario.ca/page/ministry-environment-and-climate-change-ministers-annual-report-drinking-water-2016.

16. Ministry of Transportation. 2017c. *Electric vehicle incentive program (EVIP)* [Online]. http://www.mto.gov.on.ca/english/vehicles/electric/electric-vehicle-incentive-program.shtml.

17. Ministry of Transportation. 2017b. *Electric vehicle charging incentive program* [Online]. http://www.mto.gov.on.ca/english/vehicles/electric/charging-incentive-program.shtml.

18. Ministry of Transportation. 2017a. *Electric vehicle chargers Ontario (EVCO)* [Online]. http://www.mto.gov.on.ca/english/vehicles/electric/electric-vehicle-chargers-ontario.shtml.

19. Save on Energy. 2017. *FAQs* [Online]. https://saveonenergy.ca/Business/Program-Overviews/Audit-Funding/Audit-Funding-FAQs.aspx.

20. Ministry of Energy. 2015. Helping small businesses save energy. http://www.energy.gov.on.ca/en/files/2015/10/Helping-Small-Businesses-Save-Energy.pdf.

21. Ibid.

22. Ministry of Energy. 2017c. Smart grid fund guidelines. http://www.forms.ssb.gov.on.ca/mbs/ssb/forms/ssbforms.nsf/GetFileAttach/011-0031E~2/$File/0031E_Guide.pdf.

23. Government of Ontario. 2017b. *Greener diesel regulation* [Online]. https://www.ontario.ca/page/greener-diesel-regulation.

24. Ministry of Transportation. 2017d. *Ontario's green licence plate program* [Online]. http://www.mto.gov.on.ca/english/vehicles/electric/green-licence-plate.shtml.
25. Ministry of Agriculture, F. A. R. A. 2016b. *Tax credit for farmers who donate food—Bringing more local food to communities across Ontario* [Online]. http://www.omafra.gov.on.ca/english/about/info-taxcredit.htm.
26. Greenbelt Fund. 2017b. *Local food literacy grant stream* [Online]. http://www.greenbeltfund.ca/local_food_literacy_grant_stream.
27. Greenbelt Fund. 2017a. *Broader public sector grant stream* [Online]. http://www.greenbeltfund.ca/broader_public_sector_grant.
28. Greenbelt Fund. 2017c. *Market access grant scheme* [Online]. http://www.greenbeltfund.ca/market_access_grant_stream.
29. Ministry of Agriculture, F. A. R. A. 2017a. *Growing Forward 2 for Organizations and Collaborations* [Online]. http://www.omafra.gov.on.ca/english/about/growingforward/gf2-org.htm.
30. Ministry of Agriculture, F. A. R. A. 2017b. *Growing Forward 2 for Processors* [Online]. http://www.omafra.gov.on.ca/english/about/growing-forward/gf2-processor.htm.
31. Ministry of Agriculture, F. A. R. A. 2017c. *Growing Forward 2 for Producers* [Online]. http://www.omafra.gov.on.ca/english/about/growing-forward/gf2-farmbus.htm.
32. Ministry of Agriculture, F. A. R. A. 2016a. Ontario's local food report 2015/16 edition. http://www.omafra.gov.on.ca/english/about/local_food_rpt16.pdf.

References

Government of Ontario. 2010. *Water Opportunities Act, 2010, S.O. 2010, c. 19, Sched. 1* [Online]. https://www.ontario.ca/laws/statute/S10019
———. 2016. *Large building energy and water reporting and benchmarking* [Online]. https://www.ebr.gov.on.ca/ERS-WEB-External/displaynoticecontent.do?noticeId=MTI3ODY0&statusId=MTkzMTc3
———. 2017a. *Green investment fund* [Online]. https://www.ontario.ca/page/green-investment-fund
———. 2017b. *Greener diesel regulation* [Online]. https://www.ontario.ca/page/greener-diesel-regulation

————. 2017c. *Permits to take water* [Online]. https://www.ontario.ca/page/permits-take-water

————. 2017d. *Source protection* [Online]. https://www.ontario.ca/page/source-protection

Greenbelt Fund. 2017a. *Broader public sector grant stream* [Online]. http://www.greenbeltfund.ca/broader_public_sector_grant

————. 2017b. *Local food literacy grant stream* [Online]. http://www.greenbeltfund.ca/local_food_literacy_grant_stream

————. 2017c. *Market access grant scheme* [Online]. http://www.greenbeltfund.ca/market_access_grant_stream

Invest Ontario. 2017. *Food and beverage manufacturing* [Online]. http://www.investinontario.com/food#diverse

Kubursi, A., H. Cummings, R. Macrae, and P. Kanaroglou. 2015. Dollars & sense: Opportunities to strengthen Southern Ontario's food system. http://www.greenbelt.ca/dollars_and_sense_opportunities_2015

Mentorworks. 2017. *Ontario innovation demonstration fund* [Online]. https://www.mentorworks.ca/what-we-offer/government-funding/capital-investment/idf/

Ministry of Agriculture, Food and Rural Affairs. 2016a. Ontario's local food report 2015/16 edition. http://www.omafra.gov.on.ca/english/about/local_food_rpt16.pdf

————. 2016b. *Tax credit for farmers who donate food—Bringing more local food to communities across Ontario* [Online]. http://www.omafra.gov.on.ca/english/about/info-taxcredit.htm

————. 2017a. *Growing Forward 2 for organizations and collaborations* [Online]. http://www.omafra.gov.on.ca/english/about/growingforward/gf2-org.htm

————. 2017b. *Growing Forward 2 for processors* [Online]. http://www.omafra.gov.on.ca/english/about/growingforward/gf2-processor.htm

————. 2017c. *Growing Forward 2 for producers* [Online]. http://www.omafra.gov.on.ca/english/about/growingforward/gf2-farmbus.htm

Ministry of Energy. 2015. Helping small businesses save energy. http://www.energy.gov.on.ca/en/files/2015/10/Helping-Small-Businesses-Save-Energy.pdf

————. 2017a. *Energy and water reporting and benchmarking for large buildings* [Online]. http://www.energy.gov.on.ca/en/ontarios-ewrb/

————. 2017b. *Ontario's Green Button initiative* [Online]. http://www.energy.gov.on.ca/en/ontarios-electricity-system/green-button/

———. 2017c. Smart grid fund guidelines. http://www.forms.ssb.gov.on.ca/ mbs/ssb/forms/ssbforms.nsf/GetFileAttach/011-0031E-2/$File/0031E_ Guide.pdf

Ministry of the Environment and Climate Change. 2016. *Minister's annual report on drinking water 2016* [Online]. https://www.ontario.ca/page/ ministry-environment-and-climate-change-ministers-annual-report-drinking-water-2016

Ministry of Transportation. 2017a. *Electric vehicle chargers Ontario (EVCO)* [Online]. http://www.mto.gov.on.ca/english/vehicles/electric/electric-vehicle-chargers-ontario.shtml

———. 2017b. *Electric vehicle charging incentive program* [Online]. http:// www.mto.gov.on.ca/english/vehicles/electric/charging-incentive-program. shtml

———. 2017c. *Electric vehicle incentive program (EVIP)* [Online]. http://www. mto.gov.on.ca/english/vehicles/electric/electric-vehicle-incentive-program. shtml

———. 2017d. *Ontario's green licence plate program* [Online]. http://www.mto. gov.on.ca/english/vehicles/electric/green-licence-plate.shtml

Mowat Centre. 2013. *Background report on the Ontario energy sector.* Toronto: Mowat Centre.

Ontario Energy Board Commission. 2016. Ontario energy report Q3 2016. http://www.ontarioenergyreport.ca/pdfs/5993_IESO_Q3OER2016_ Electricity_EN.pdf

Ontario Ministry of Finance. 2017a. Economic trends and projections. *Ontario's Long-Term Report on the Economy* [Online]. http://www.fin.gov.on.ca/en/ economy/ltr/

———. 2017b. Ontario population projections update, 2015–2041.

Save on Energy. 2014. Audit funding. https://saveonenergy.ca/Business/ Program-Overviews/Audit-Funding.aspx

———. 2017. *FAQs* [Online]. https://saveonenergy.ca/Business/Program-Overviews/Audit-Funding/Audit-Funding-FAQs.aspx

WaterTAP. 2017. *The story* [Online]. http://watertapontario.com/the-story/

8

The Green Economy and the Water-Energy-Food Nexus in Denmark

Introduction

Since the mid-1990s, Denmark's growth in productivity has been relatively weak compared to other OECD countries. The result has been that Denmark's annual GDP growth over the past 20 years has only been 1.6 percent.[1] Economic growth is projected to gradually strengthen to 1.9 percent in 2018 due to investments and exports.[2] Denmark currently has a population of 5.7 million.[3] By 2030, its population is projected to increase to 6.1 million, and by 2060, it will reach 6.4 million.[4]

8.1 Water-Energy-Food Nexus Pressures

Denmark is experiencing a variety of water, energy, and food nexus pressures that are detrimental to the development of a green economy as described below through a variety of examples.

© The Author(s) 2018
R.C. Brears, *The Green Economy and the Water-Energy-Food Nexus*,
DOI 10.1057/978-1-137-58365-9_8

Water

The Danish drinking water supply is entirely based on groundwater, and the government's official position is that drinking water should be based on pure groundwater, which only needs simple treatment before it is distributed to consumers. While groundwater quality in deeper aquifers is generally good, removing the need for expensive water purification, many shallow aquifers suffer from groundwater pollution especially from nitrates and pesticides. Even in suburban areas, nitrate pollution of groundwater has led to closures of many minor water works reliant on shallow aquifers, while pesticide pollution has led to closures of several major well fields. Meanwhile, leaching from waste disposals and other point sources has closed several water supply wells. Over the period 1991–2005, over 1300 wells were closed as water supply abstraction wells, solely due to content of pesticides or degradation. Even today, around 100 wells are closed every year due to pesticide content.[5] Climate change will pose a challenge to water quality in Denmark with more frequent and severe rainfall events likely to cause flooding in cities and pollute groundwater supplies.[6]

Energy

Electricity consumption in Denmark has increased by around 13 percent from 1990 to 2014. Forecasts project an increase in electricity consumption of 7.8 TWh from 2015 to 2025, corresponding to an increase of 23 percent. Part of this increase in demand is due to the introduction of electric boilers, heat pumps, and electric vehicles. Another factor in the increase in electricity consumption is the installations of several large data centres in Denmark, which are expected to have an electricity consumption corresponding to over 13 percent of the total consumption in 2025.[7] Nonetheless, electricity from renewables accounts for 53.4 percent of Danish domestic electricity supply. Of this, wind power accounts for nearly 38.8 percent; biomass accounts for 11.4 percent; and solar energy, hydro, and biogas account for the remaining 3.2 percent.[8]

Food

Denmark's total agricultural area is 2.7 million hectares, which amount to 61 percent of Denmark's total surface area. The country produces an amount of food sufficient to supply 15 million people every year, which is three times the Danish population.[9] The main crops in Denmark are small grains, mainly wheat and barley, covering more than half of the agricultural area. Over time, the number of farms has become fewer and larger with their productivity and profitability increasing. However, intensification in agricultural production has led to significant pollution of waterways as well as coastal water with nutrients and organic matter. Nonetheless, around 7 percent of Denmark's total agricultural area is cultivated organically with the largest organic sector being vegetables with organic products corresponding to around 20 percent of the total vegetable production.[10]

8.2   General Fiscal Tools to Reduce Water-Energy-Food Nexus Pressures

Denmark has implemented a variety of fiscal tools that create interdependencies and synergies between the nexus systems while reducing trade-offs between the systems in the development of a green economy.

Fund for Green Business Development

The Fund for Green Business Development promotes resource efficiency in Danish businesses by giving grants to selected businesses for the financing of new green products, sustainable materials, or green services. Each project will receive funding of up to DKK 1 million per project and contribution to project costs from the fund recipients will be as follows:

- Small businesses: Up to 60 percent
- Medium-sized businesses: Up to 50 percent
- Large companies: Up to 40 percent
- Public institutions: Up to 50 percent.

The Fund supports businesses, organisations, and partnerships wishing to apply for grants for projects that fall under the following six themes:

1. Development of new green business models
2. Product innovation and re-design of products
3. Promotion of sustainable materials in product design
4. Sustainable transition in the textile and fashion industry
5. Reducing food waste
6. Sustainable bio-based products based on non-food biomass

The selected projects all feature a high degree of novelty, have the potential to create growth and new green jobs in Denmark, and lead to environmental improvements.[11]

Green Business Models

The Fund for Green Business Development, the Danish Regions, and the Regional Municipality of Bornholm have partnered to establish an accelerator programme on green business model innovation. The programme aims to promote the development of new green business models by Danish businesses to generate growth and employment for the businesses and reduce their consumption of resources and environmental impact at the same time. The programme has two phases:

1. Phase one: In phase one, businesses can apply for support to develop an overall business plan describing the concept and future cash flows and presenting a detailed business plan for the green business model.
2. Phase two: In phase two, businesses can receive support for testing and executing the model in its activities.

In phase one, the Fund for Green Business Development supports businesses in their initial phase of the accelerator programme with up to DKK 250,000 to develop their overall business plan. In the second phase the Danish Regions and the Regional Municipality provide up to DKK

1,000,000 in support with the regional support linked to the business's geographical location.[12]

8.3   General Non-fiscal Tools to Reduce Water-Energy-Food Nexus Pressures

Denmark has implemented a variety of non-fiscal tools that create inter-dependencies and synergies between the nexus systems while reducing trade-offs between the systems in the development of a green economy.

The Partnership for Green Public Procurement

The Partnership for Green Public Procurement (GPP) was established in 2006 by the Ministry of Environment and Denmark's three largest municipalities (Copenhagen, Aarhus, and Odense) with the objective of creating a coalition of government bodies to affect a larger procurement volume and therefore have a larger impact on the market. Since then the partnership has been joined by a total of nine municipalities, two regions, and a water supply company. The partnership in total represents around EUR 5.5 billion a year, around 13 percent of the annual public procurement in Denmark and 30 percent of the municipalities' total procurement volume.

Forum for Sustainable Procurement

A complementary initiative of the GPP is the Forum for Sustainable Procurement to promote environmentally conscious and responsible procurement of goods and services among professional procurers in both the public and private sectors. Established in 2011, the forum acts as a 'knowledge hub' for green procurement by disseminating knowledge and facilitating experience sharing through conferences, seminars, newsletters, network groups, case studies, and web-based communication. Currently the forum has 800 members. The GPP is based on joint, mandatory green procurement criteria and targets. For example, in the area of food, the GPP's criteria include the requirement of purchasing seasonal fruit and vegetables; a minimum of 60 percent of coffee, tea, and sugar should be organic; and the mapping of organic food.[13]

Green Industrial Symbiosis Denmark

In 2012, the Danish government established a national industrial symbiosis programme, Green Industrial Symbiosis Denmark, administered by the Danish Business Authority, an agency under the Danish Ministry for Business and Growth. The aim of the programme is to promote competitiveness and resource efficiency through the establishment of commercial partnerships between companies that can make use of each other's excess resources, for example, waste materials, wastewater, or surplus heat. Businesses that enter into an industrial symbiosis with one or more other businesses can receive numerous benefits including reduced costs in disposing of wastewater; revenue from selling waste; lower costs for basic commodities and raw materials if purchased as residual materials from other businesses; reduced expenditure on energy and water; and an enhanced 'greener' profile from reducing consumption of new resources and lowering of carbon emissions.[14]

Taskforce for Resource Efficiency

In 2014, the Danish government, in partnership with the five Danish regions, the Regional Municipality of Bornholm, and the Danish Business Authority, established the Taskforce for Resource Efficiency. The taskforce is comprised of technical experts that offer free resource checks to companies and facilitate matchmaking. The Taskforce also provides assistance in drafting action plans and applying for subsidies.[15] In 2015, the taskforce began identifying regulatory barriers to increased resource and water efficiency with the goal of setting up a solutions team in 2016–2017 to find the most effective ways of overcoming these barriers.

Danish Nature Interpretation Service

Since 1987, Denmark has built up a Nature Interpretation Service which inspires people to take care of nature, be aware of the climate and the good health of people, and have a positive attitude towards sustainable development. The Service consists of a network of 360 nature interpreters and their employees that operate in close cooperation between the Danish Ministry of the Environment and the Danish Outdoor Council. The nature interpreters are positioned in municipalities, museums, visitor centres, and 'green' organisations. The Service works with a wide range of target groups including kindergarten and school children and adults with various educational backgrounds. The national network overall gives people experience in nature with the aim of:

- Strengthening their understanding of nature, biodiversity, environment, heritage, and the necessity of sustainable development
- Strengthening their outdoor possibilities and activities
- Giving them inspiration to participate in a democratic process during involvement in the management of the natural and cultural environment
- Giving them inspiration for a more healthy and sustainable lifestyle[16]

8.4   Water: Non-fiscal Tools

Denmark has implemented a variety of water-related non-fiscal tools that create interdependencies and synergies between the nexus systems while reducing trade-offs between the systems in the development of a green economy.

Pesticides Strategy to Protect Waterways

Since 2007, the pesticide load in Denmark's waterways has increased by 35 percent. In response the Danish government launched its Pesticides Strategy that aims to reduce the pesticide load by 40 percent by the end of

2015 compared with 2011, ensuring a clean environment, good ecological conditions in nature, healthy food, better health and safety at work, and more green workplaces. As part of this strategy the government has developed a series of benchmarks including:

- No exceedance of the threshold value for groundwater for approved pesticides.
- The content of pesticide residues in Danish food must be as low as possible.
- Human health loads from using substances of very high concern (carcinogenic and endocrine disruptors) must be reduced by 40 percent by the end of 2015 compared to 2011.
- Anyone that uses pesticides for professional purposes must comply with the principles of integrated pest management (IPM) by 2014.
- Municipalities, regions, and government institutions must reduce their use of pesticides and pesticide loads in accordance with the voluntary agreement in this area, golf courses must reduce their pesticide use and loads, and owners of private gardens must choose pesticides causing low loads.[17]

Focus on Integrated Pest Management

IPM means that farmers must do everything possible to prevent attacks of pests and thereby reduce the need for pesticides. Prevention and control of pests can be achieved by having, for example, a varied and healthy crop rotation; using where possible resistant or tolerant crop types; and reducing pesticide application frequency or partial application to prevent pests from developing resistance. A new pesticide tax will be implemented to promote the use of pesticides resulting in the lowest load on human health, nature, and groundwater. Pesticides causing the highest loads will be taxed more heavily to encourage farmers and other growers to comply with IPM principles, and use less pesticides and use pesticides causing the lowest load. In addition to the tax the Danish government will enhance educational initiatives on IPM (Table 8.1).[18]

Table 8.1 Danish Government educational initiatives on integrated pest management

Initiative	Description	Effect
Targeted IPM advisory services in agriculture	The most recent knowledge on IPM will be tested on special IPM demonstration farms with knowledge conveyed to the entire agricultural industry.	Reduction in the use of pesticides and reduced load
Targeted IPM advisory services in horticulture	To reduce pesticide residues in fruit and vegetables and impacts on health and safety in greenhouse production, the most recent knowledge will be tested on demonstration nurseries testing IPM principles with knowledge conveyed to the horticultural industry.	Reduction in the use of pesticides and reduced load
Advisory services on minimising pesticide residue in food	Advisory services will focus on ensuring the lowest possible level of pesticide residue in crops.	Fewer pesticide residues in food
Updating cultivation guidelines	Crop-specific cultivation guidelines for agriculture/nurseries will be updated to include IPM tools considered most realistic to implement.	Reduction in the use of pesticides and reduced pesticide loads
Increasing focus on IPM in the education of farmers and advisory	Knowledge about IPM will become part of the curriculum in a pesticide application certificate training programme, and agricultural advisors will be offered courses on IPM.	Reduction in the use of pesticides and reduced pesticide loads

Reduced Pesticide Usage in Public Areas

The pesticides strategy aims to phase out pesticides in public areas with a focus on phasing out pesticides in parks and sports areas. In 2007, government institutions, regions, and municipalities renewed an agreement from 1998 to phase out the use of pesticides in public areas. To date there has been a significant decrease in the use of pesticides in public areas from 23 tonnes in 1995 to three tonnes in 2010. To reduce pesticide use

Table 8.2 Reducing pesticide usage in public areas

Initiative	Description	Effect
Pest control strategies for invasive plant species	Pest control strategies will be updated regularly with the most recent knowledge and experience gained by municipalities, regions, and government institutions.	Efficient control of invasive plant species and maintaining low use of pesticides
Increased knowledge sharing between municipalities and governmental institutions	A forum will be established for municipalities and government institutions to exchange experience on managing public areas without pesticides.	Maintaining a low use of pesticides
Survey of the use of pesticides	Data on the use of pesticides in public areas during 2013 will be published.	Exchange of information and overview of developments

further, the strategy aims to develop pest control strategies for controlling invasive species, increase knowledge sharing between municipalities and governmental institutions, and survey the use of pesticides (Table 8.2).[19]

Strict Demands on Golf Courses

The pesticides strategy aims to reduce pesticide use and load by golf courses, which use around two tonnes of pesticides annually. The Danish government has introduced a cap on the amount of pesticide that can be used on individual golf courses. In future, annual use of pesticides must be reported to the Danish Environmental Protection Agency (EPA) with the Agency monitoring compliance with the new cap. These requirements will be supplemented by initiatives that increase knowledge and change attitudes in the use of pesticides. The strategy aims to enhance training of greenkeepers, create an Internet portal for reporting pesticide use, continue inspection campaigns of golf courses, and evaluate regularly the new requirements (Table 8.3).[20]

Table 8.3 Initiatives to reducing pesticide usage on golf courses

Initiative	Description	Effect
Enhance training of greenkeepers	Greenkeepers' level of knowledge needs increasing through training courses that focus on alternative pest control strategies, management of golf courses without pesticides, tailored dosages, and calibration and maintenance of spraying equipment.	Reduced pesticide use and pesticide load
New Internet portal for reporting pesticide use	A reporting portal will be established for golf clubs to report their use of pesticides, enabling the monitoring of usage, calculation of loads on individual golf courses, and monitoring of compliance with regulations.	Reduced pesticide use and pesticide load
Continuation of inspection campaigns on golf courses	The Danish EPA will carry out inspection campaigns on golf courses to check on the usage of pesticides in the new reporting portal.	Reduced pesticide use and pesticide load
Evaluation of new regulations	Regulations on the maximum pesticide load on golf courses will be evaluated after the first year and afterwards once every three years and whether it is necessary to tighten regulations further.	Reduced pesticide use and pesticide load

Reduced Pesticide Load in Private Gardens

Owners of private gardens use around 60–120 tonnes of pesticides per annum to control moss and weeds, corresponding to around 2 percent of total pesticide sales in Denmark. The pesticides strategy aims to increase the public's knowledge on the risks of using individual pesticides and encourage owners of private gardens to either stop using pesticides completely or choose products with the lowest load on human health, nature, and groundwater. In addition, distributors will be better equipped to advise customers and pesticides can be purchased only via direct contact with a salesperson (Table 8.4).[21]

Table 8.4 Initiatives to reducing pesticide loads in private gardens

Initiative	Description	Effect
Information campaign on choice of pesticide	Information campaigns will provide owners with tools to choose pesticides causing the lowest load and to use these safely. Information campaigns will increase knowledge about alternatives to pesticides.	Increased protection of human health, nature, and ground water
Information campaign on illegal imports of pesticides	An information campaign will be carried out to inform owners that only pesticides approved by the EPA are allowed.	Increased protection of human health, nature, and ground water
More restricted access to purchasing pesticides	Criteria will be developed to determine which pesticides can be purchased and used by persons without a spraying certificate (i.e. owners of private gardens). This will ensure only pesticides with the lowest load can be purchased for private gardens.	Increased protection of human health, nature, and ground water
New terms for selling pesticides to owners of private gardens	Regulations will be introduced to ensure salespersons have a certificate to sell pesticides. This means the majority of pesticides will be kept behind the counter, in storage or in locked cabinets/rooms in the shop.	Increased protection of human health, nature, and ground water
Annual statistics of sales to owners of private gardens	Annual sales statistics will be produced to provide knowledge about which pesticides private garden owners buy.	Better knowledge from more targeted information

Adapting to Water-Related Climate Change Extreme Events

Adapting to extreme weather events (e.g. floods and droughts) requires collaboration between authorities, organisations, private enterprises, and individuals irrespective of whether the project is related to maintenance of roads, coastal protection, construction, or investments in new infrastructure. The responsibility of the central government in Denmark is to establish a framework for local climate change adaptation by adapting laws and regulations and ensuring coordination and providing of information. This framework enables local municipalities to incorporate climate change adaptation into local master plans and that businesses and individuals are able to prevent damage to buildings, avoid health risks, and exploit economic opportunities from climate change adaptation.[22] A variety of initiatives have been undertaken to enhance resilience to climate change or extreme weather events including floods and droughts as follows:

Online Vulnerability Mapping

The Ministry of Housing, Urban and Rural Affairs will create an online tool for municipalities and property owners to map vulnerability to climate change of local areas and buildings.[23]

Danish Portal for Climate Change

The Ministry of the Environment has developed the Danish Portal for Climate Change to advise farmers and companies about prevention of the negative impacts of climate change in relation to property.[24]

Public–Private Collaboration on Climate Change Adaptation

Under the Urban Renewal Act the public sector leads the way in investments to initiate economic development in urban areas. To obtain state funding municipalities have to cooperate with the private sector with climate change adaptation being one of the collaborative themes.[25]

Sustainable Land Development and Climate Change Adaptation Certification

The Ministry of the Environment, in collaboration with the Danish National Association for Building, the Confederation of Danish Industries, and Green Building Council of Denmark, is testing a certification scheme for sustainable land development in a number of Danish construction projects. The criteria for sustainability include a focus on climate change adaptation.[26]

State of Green's Rethink Water Network

The Rethink Water network comprises over 60 companies, organisations, institutions along with government bodies specialising in water efficiency. The network brings together a diverse mix of consulting companies, technology suppliers, clients, researchers, and governmental bodies to share knowledge and through collaboration create solutions for water security and increased water efficiency.[27]

8.5   Energy: Fiscal Tools

Denmark has implemented a variety of energy-related fiscal tools that create interdependencies and synergies between the nexus systems while reducing trade-offs between the systems in the development of a green economy.

Taxes to Encourage Energy Efficiency

Since 1977, successive Danish governments have been using energy taxes to encourage energy efficiency. Over the years the scope of energy taxation has widened and rates gone up resulting in Danish energy taxes being among the highest in the world: the average Danish consumer price per kWh is almost EUR 0.30, of which EUR 0.17 is taxes.[28]

8.6   Energy: Non-fiscal Tools

Denmark has implemented a variety of energy-related non-fiscal tools that create interdependencies and synergies between the nexus systems while reducing trade-offs between the systems in the development of a green economy.

Voluntary Agreement Scheme on Energy Efficiency

The Danish voluntary agreement scheme on energy efficiency for energy-intensive industries was launched in 1996. The scheme was launched by the Danish government to achieve the long-term goal of reducing greenhouse gases by 20 percent in 2005 compared to 1988. By 2010 around 230 companies had voluntary agreements covering about two thirds of total fuel consumption. The scheme is based on the principle that while it is voluntary companies entering the scheme will have economic incentives to encourage efficiencies to be made. The Danish government has chosen the immediate economic benefit of participating being an energy tax relief. As such, the voluntary scheme is designed for companies that operate certain energy-intensive operations or for those companies whose energy taxes exceed 4 percent of the companies' value added.[29]

Requirements to Participate in the Voluntary Scheme

A voluntary energy efficiency agreement is a three-year contract entered between the company and the Danish Energy Agency. It is also possible that a number of companies in the same industrial sub-sector enter into a joint agreement with the Danish Energy Agency where a joint agreement consists of individual agreements and a number of special investigations relevant to the sub-sector. The main requirements of companies entering the scheme is the company must implement and maintain a certified energy management system according to the global standard ISO 50001; the company must carry out special investigations and projects that focus on their primary production processes including a thorough productivity and optimisation analysis; and the company must implement all energy efficiency projects with a payback period of four years or less.

Annual Certification of the Voluntary Agreement

The energy management system and the additional requirements must then be certified annually by an accredited body verifying the company:

- Presents an updated annual breakdown of energy consumption by end-use
- Sets targets and budgets for energy consumption in the coming year
- Screens the company to identify possible energy-saving projects
- Prepares an action plan for investments in the coming year
- Implements all energy-saving measures and projects with payback periods of four years or less
- Carries out special investigations for complicated energy-saving areas
- Evaluates energy key performance indicators regularly
- Applies energy-efficient design methods when planning investments
- Carries out internal audits of procedures and reporting
- Arranges management evaluation of the scheme
- Follows procedures for energy-efficient design and purchasing

Overall the costs of the scheme, including internal time allocated to managing the scheme, certification costs, and fees for external specialists for most energy-intensive companies, are less than the energy tax relief they receive. However, for SMEs a certified energy management system is generally too expensive to maintain and therefore the Danish Energy Agency with the Danish Standards (accredited to certify energy management systems) has introduced an 'energy management-light' methodology for this target group.[30]

Energy Management-Light for SMEs

The energy management-light system is not a certified scheme but recommends an organisation to address the cornerstones of an efficient energy management system, in particular define a target for the energy-saving effort; organise the right team for the effort; map energy consumption; create an action plan; establish relevant key energy performance indicators; have management evaluate progress regularly; and establish simple annual procedures to follow. In addition to SMEs, energy management-light principles have been adopted by a variety of energy users including municipal buildings, large offices, and hospitals.[31]

Working with Industry to Create an Evolving Relevant Scheme

To ensure the success of the voluntary scheme the Danish Energy Agency has a small, specialised team to administer and develop the scheme. While annual control of the agreements has been outsourced to external, accredited companies within energy management systems, the Danish Energy Agency has its own resources to adjust agreements as well as review progress reports from the companies. This ensures the Danish Energy Agency continuously understands the progress and barriers towards the success of the scheme. The Danish Energy Agency also has close dialogue with industry to understand priorities and new agendas for integrating energy efficiency activities within company life. At times an industrial advisory board, with experts from relevant industries, has been formed to work with the Danish Energy Agency in developing the agreement scheme into new areas and ways of working.[32]

Supportive Measures

The Danish Energy Agency has initiated a range of supportive measures to develop and secure the success of the voluntary scheme. Measures include developing:

- A consultant scheme to support industries with specialist competencies within the energy efficiency of industrial processes and utility systems
- Fact sheets and case studies to showcase and demonstrate projects
- A 'toolbox' development project managed by industry people that demonstrates tools and methods within energy management systems and energy performance indicators

- A range of surveys and assessments to establish a database for industrial energy use, for example, surveys of energy-saving potentials,
review of experiences from special investigations, and analysis of
potential uses of renewable energy[33]

Compulsory Energy Audits

In 2014, the Danish Parliament passed an Act requiring large enterprises
to carry out a mandatory energy audit every fourth year. The enterprise
can also satisfy the obligation by using and maintaining a certified energy
or environmental management system that includes an energy audit.
Under this Act, 'large enterprises' mean enterprises with at least 250
employees and an annual turnover of more than EUR 50 million or an
annual balance sheet of more than EUR 34 million.[34]

Better Homes

BedreBolig (Better Homes) is a scheme introduced in 2014 that makes
it easier for homeowners to renovate their homes by offering comprehensive, expert advice throughout the energy renovation process. Better
Homes focuses on developing cooperation between homeowners and
financial institutions. The aim is to make information more easily available to banks and mortgage institutions so they can advise customers on
the financing of energy improvements.[35]

Energy-Labelling Scheme for Buildings

The energy-labelling scheme for buildings means that all buildings in
which energy is used to regulate the indoor climate must be energy-
labelled before they are sold or let. This involves an inspection of the
building by an independent expert who rates the energy efficiency of the
building on a scale from A2020 to G, where A2020 is the best. It also
covers the potential energy savings that can be made to the building.
Since 2010, it is mandatory that energy-ratings are to be included in

all advertisements for sale of properties through real estate agents, helping create transparency on the property market as studies have shown a relationship between property prices and energy certificates. In 2013 this requirement was extended with the requirement that the energy label for buildings must now be included in all advertisements for the sale or letting of buildings, flats, or commercial premises.[36] Due to the large volume of data being collected on the design and energy characteristics of buildings the Danish Energy Agency has created a web portal where tradespeople, construction companies, and research and development institutes can have access to a larger knowledge base on the design and energy status of buildings, facilitating the development of new and innovative energy-efficient solutions and services. The Danish Energy Agency has also developed a website for building owners to analyse potential energy savings for single-family houses on the basis of data collected from energy labelling. In the future this website will be expanded to enable building owners to calculate the financing costs of energy renovation projects.[37]

8.7 Food: Fiscal Tools

Denmark has implemented a variety of food-related fiscal tools that create interdependencies and synergies between the nexus systems while reducing trade-offs between the systems in the development of a green economy.

Organic Land Subsidy Scheme

Denmark's Organic Land Subsidy Scheme provides a subsidy of almost EUR 117 per hectare a year for conversion, preservation, and sustainability of organic farms. In addition, the government provides a supplement of EUR 161 per hectare a year for land areas under conversion to organic production, for the first two years of conversion. The government introduced from 2015 a supplement of more than EUR 537 per hectare a year for organic areas used for fruit and berry production. To reduce the impacts of agricultural production on water resources, the government introduced from 2015 a supplement of EUR 67 per hectare a year for organic areas with reduced use of nitrogen. At the same time the government has appointed a working group comprising authorities, the industry, and organisations to cooperate on increasing organic producers' access to plant nutrients including source-separated household waste and organic waste from the service sector.[38]

Green Development and Demonstration Programme Grant Scheme

In 2010, the Danish Ministry of Food, Agriculture and Fisheries established the Green Development and Demonstration Programme (GUDP), a modern grant scheme designed to enhance competitive pressures and encourage enterprises to renew themselves strategically in order to develop new environmental technology and know-how in the creation of a sustainable, green Danish food sector. Unlike traditional public grants that are typically provided to businesses to support selected and successful enterprises, GUDP has been designed to motivate food businesses, researchers, and farmers to cooperate in order to ensure growth and at the same time address societal challenges including climate change and environmental degradation. As such, GUDP determines the problem needing to be solved and invites applicants to find the required solutions.[39]

The competition for the DKK 200 million available from GUDP is tough and the requirements stringent, involving a four-step process:

1. GUDP evaluates the project proposals against several criteria including socio-economic impacts and growth opportunities as well as the project's expected contribution to enhanced sustainability, for example, reduced nutrient leakage or pesticide usage.
2. The projects are compared to existing practices in the sector.
3. GUDP evaluates the project's uncertainty in relation to obtaining success against its potential impacts and the quality of its business plan.
4. GUDP reviews the cost effectiveness of the project: the rate of return in terms of expected impacts per DKK invested by GUDP.

The important aspect of each project is having a business plan that documents the desired outcomes and impacts achieved, not in the form of reports or papers, but in the form of innovative and concrete products, novel processes, or new knowledge, which are commercially viable and may be subsequently marketed and sold to consumers or other businesses. In particular, the business plan should describe the desired product, the business concept, the market situation, and relevant risk factors. With a business plan applicants show how they intend to turn outcomes of the project into concrete, commercial actions. The business plan forms the basis for an evaluation of the expected returns of the project.[40]

Overall, GUDP only provides grants to projects that contribute to the development of the food sector of tomorrow based on interactions between sustainability, efficiency, and value enhancement. In 2011, DKK 195 million was granted to 46 projects. The selected projects cut carbon emissions by 780,000 tonnes, reduced nutrients (10,500 tonnes of nitrogen and 1000 tonnes of phosphorus), and produced an overall economic impact of DKK 1.5 billion.[41]

Fertiliser Accounts in Denmark

In Denmark, farmers must enter the Register for Fertilizer Account if their annual turnover relating to agricultural activity is more than DKK 50,000 (EUR 6600) and they meet at least one of the following conditions of:

- Having more than 10 livestock units
- Having more than 1.0 livestock unit per hectare
- Receiving more than 25 tonnes of livestock manure

Farmers who enter the Register are required to prepare a fertiliser plan and keep it for five years, calculate the nitrogen quota for the farm, and submit a fertiliser account. The fertiliser account contains information about:

- Area sizes and type of crops (the area size of the farm is the sum of the cultivated, uncultivated, and set-aside areas)

- The nitrogen standard for the crop (all crops are given a nitrogen standard)
- The calculated nitrogen quota for the farm (the nitrogen quota of the farm is the sum of nitrogen quota of each field, where the field nitrogen quota is calculated on the basis of the size and the nitrogen standard), and the overall nitrogen quota of the farm provides the amount of fertiliser (manure and chemical fertiliser) that can be applied on the farm)
- Number of livestock units and type of livestock (animal type and number and the type of housing, feedstuffs, production, etc., so that the amount of nitrogen in the manure produced can be calculated)
- Use of fertilisers (both livestock manure and chemical fertiliser)
- Delivery of chemical fertiliser (farmers must report the amount and type of fertiliser supplied)
- Exchange of fertiliser or manure (farmers can exchange fertiliser to other farmers who are in the Register)
- Manure and fertiliser stock (opening and closing stock for the growing season should be calculated annually)[42]

Each year the Danish AgriFish Agency will visit around 1 percent of the farms and an administrative control is run on around 4 percent of the farmers that submit a fertiliser account. Farmers who are registered are then allowed to buy chemical fertiliser without paying tax on fertiliser (EUR 0.66 per kg of nitrogen). Meanwhile, farmers with an annual turnover between DKK 20,000 (EUR 2600) and DKK 50,000 may voluntarily enter the Register. Overall, this is part of Denmark's aim of reducing nitrogen leaching by 13 percent in 2015 compared to 2003 levels and reducing phosphorous surplus by 50 percent over the period 2002–2015.[43]

Alternative Plant Protection Product Grants

As part of Denmark's Pesticides Strategy, the government has set aside DKK 1 million for 2016 to assist applicants seeking authorisation for placing on the market their alternative plant protection products. As part of the Alternative Plant Protection Product Grants programme, applicants develop an application for approval of a basic substance or seek reimbursement for an application fee already paid to get an authorisation. The grant for alternative products can be up to 60 percent of eligible costs; however, for SMEs, grants can be awarded for up to 80 percent of eligible costs.[44]

Biogas from Agricultural Waste Tax Exemption

The Danish government is aiming for 50 percent of livestock manure to be made into biomass. To promote its development biogas is currently exempt from Danish taxation with the goal of encouraging the development of 50 large-scale biogas plants in Denmark.[45]

8.8   Food: Non-fiscal Tools

Denmark has implemented a variety of food-related non-fiscal tools that create interdependencies and synergies between the nexus systems while reducing trade-offs between the systems in the development of a green economy.

Danish Organic Logo

In 1987, Denmark published the first organic legislation and launched a logo as a national organic control in 1989. The logo may be applied only on products from farms authorised for organic production. The Danish Organic Farming division monitors organic farms' compliance with the rules applicable to both organic plant and animal husbandry. The division also inspects organic suppliers producing or marketing organic feed, seeds, cereals, fertilisers, and other non-food products.

Certificate of Organic Production

As part of the organic logo programme, the certificate of organic production is a statement regarding the organic production unit of a farm and includes the last organic inspection date, and overview of conversion dates and the organic status of the farm's crops and livestock. The certificate is created based on the information that is annually reported in the organic section of the online register that all farmers in Denmark use. The AgriFish Agency will issue a certificate of organic production only when all the relevant information has been entered into the online register and been certified by the Agency. All organic producers are inspected each year by the AgriFish Agency. In addition to regular annual inspections of around 3300 organic producers and 130 organic suppliers in Denmark, the Agency also carries out random inspections based on production and risk analyses.[46]

Organic Labelling for Catering

In 2009, the Ministry of Food Agriculture and Fisheries introduced labels for the marketing of organic food products in large-scale kitchens, restaurants, cafes, hospitals, schools, and larger businesses. Large-scale kitchens can use one of three organic labels that show how large the amount of raw materials used is organic. The share is given in percentage intervals: 30–60 percent, 60–90 percent, or 90–100 percent.

In 2016, more than 1500 eating places used this label, with 248 using 90–100 percent organic raw materials in all meals served.[47]

8.9   Case Study Summary

Denmark has implemented a variety of initiatives to create interdependencies and synergies between the nexus systems while reducing trade-offs between the systems in the development of a green economy.

General

- Denmark's Fund for Green Business Development provides grants to businesses as well as public institutions that wish to implement projects that green their products, materials, or services. In addition, the fund provides grants to businesses to green their business models.
- Denmark's national government, along with municipalities, regions, and a utility, has formed a GPP coalition that creates economies-of-

scale in purchasing green, sustainable products. In addition, the coalition has created a forum that raises awareness among public and private sector procurers on sustainable products, with members having to meet sustainability purchasing criteria.

- Denmark has created a green industrial symbiosis that fosters commercial partnerships between companies that can make use of each other's excess resources and waste.
- Denmark established a Taskforce for Resource Efficiency to identify regulatory barriers to increased resource efficiency.
- In order to raise awareness of environmental sustainability, Denmark has a long-established Nature Interpretation Service that raises awareness on sustainable development.

Water

- Denmark has established a pesticides strategy to reduce groundwater pollution as well as reduce pesticide residue in food. As part of the strategy, Denmark is encouraging farmers to use IPM strategies. Furthermore, the strategy aims to phase out the use of pesticides in public areas as well as encourage private gardens to reduce their usage of pesticides.
- The Danish government has implemented a variety of measures to enhance resilience to climatic extremes that impact water resources; for instance, an online tool for municipalities and property owners to map their vulnerability to climate change in local areas and buildings will be developed.
- A sustainable land development certificate will be developed with the construction industry with a particular focus on climate change adaptation.

Energy

- Denmark has long had energy taxes on fossil fuels to encourage energy efficiency.
- To decrease the price gap between fossil fuels and renewable energy, Denmark is providing an investment subsidy for industries to convert to renewable energy as well as make significant energy efficiency improvements.

- Denmark's voluntary agreement on energy efficiency involves large companies, or a group of companies from the same industrial sub-sector, entering into a partnership with the government to implement energy management systems, have an annual audit, and make energy efficiency improvements, with participating companies receiving a tax relief. In addition, the government continuously reviews the success of the scheme by having close dialogue with industry to understand evolving priorities and new agendas for integrating energy efficiency into industrial processes.
- Denmark operates an energy-labelling scheme for buildings, ensuring all buildings which use energy to regulate the indoor climate must be energy-labelled before they are sold or let.

Food

- Denmark's organic land subsidy scheme promotes the conversion, preservation, and sustainability of organic farms. The scheme will also provide a supplement to reduce the use of nitrogen on organic areas.
- Denmark's GUDP is providing grants for food industry–related busi-nesses, researchers, and farmers to cooperate on enhancing productiv-ity while addressing environmental degradation.
- All farms with a particular annual income will be required to enter the register for fertiliser usage with participants having to have a fertilise plan with registered farms able to buy chemical fertiliser without pay-ing fertiliser tax.
- The Danish Organic label is issued by the Danish government for pro-duce that originates from farms certified as being organic. In addition, Denmark has created organic labels for large-scale kitchens, restau-rants, cafes, and so forth that show how large the amount of raw mate-rial is organic.

Overall, Denmark uses a variety of fiscal and non-fiscal tools to create interdependencies and synergies between the nexus systems while reduc-ing trade-offs between the systems in the development of a green econ-omy. These tools are summarised in Table 8.5.

Table 8.5 Denmark case summary

Tool	Tool type	Policy title	Description	WEF sectors addressed
Fiscal	Environmental taxes	Fertiliser Account	Farmers who have a registered fertiliser account are allowed to buy chemical fertiliser without paying tax on fertiliser.	Food, water
		Tax to encourage Energy efficiency	Danish energy taxes have widened over time and gone up to encourage energy efficiency.	Energy
		Biogas from agricultural waste tax exemption	The tax supports the government's aim of having 50 percent of livestock waste turned into biomass.	Food, energy
	Financial incentives	Fund for Green Business Development	Grants are given to business for the financing of new green products, materials, services, or business models.	Water, energy, food
		Investment subsidy for renewables and energy efficiency	A subsidy was established to bridge the price gap between renewable and fossil fuels and promote energy-efficiency and use of renewable energy in industrial production processes.	Energy, food
		Green Development and Demonstration Programme Grant Scheme	The scheme encourages the development of new environmental technology and know-how in the creation of a sustainable green Danish food sector.	Food, water, energy
		Alternative plant protection product grant	Alternative plant protection products can receive a grant to reduce the costs of placing the product on the market.	Food, water
	Payment for ecosystem services	Organic Land Subsidy Scheme	The scheme will provide a subsidy for conversion, preservation, and sustainability of organic farms. The government will provide a supplement for land areas under conversion to organic production and to organic areas with reduced use of nitrogen.	Food, water

(continued)

Table 8.5 (continued)

Tool	Tool type	Policy title	Description	WEF sectors addressed
Non-fiscal	Regulations	Taskforce for Resource Efficiency	The Taskforce identified regulatory barriers to increased resource and water efficiency, from which regulatory solutions will be developed.	Water, energy, food
		Pesticides strategy	The government aims to reduce pesticide loads in waterways by encouraging IPM, reducing usage in public areas and golf courses, encouraging a reduction in usage in private gardens.	Water, food
	Standards and mandatory labelling	Compulsory energy audits	All large enterprises must carry out an energy audit every four years.	Energy
		Energy labelling of buildings	All buildings which use energy to regulate the indoor climate must be energy-labelled before they are sold or let.	Energy
	Public–private partnerships	Urban Renewable Act	To obtain state funding for climate adaptation municipalities have to cooperate with the private sector.	Water
	Information and awareness-raising	Danish Portal for Climate Change	The Internet portal advises farmers and companies about adapting to climate change in relation to their property.	Food, water
		Better Homes	Expert advice on energy renovations is provided by financial institutions to homeowners.	Energy
	School education	Danish Nature Interpretation Service	Nature interpreters educate school children and adults on nature and sustainable development.	Water, energy, food
	Voluntary labelling	Certification scheme for sustainable land development	The government, along with the private sector, is testing a certification scheme on sustainable land development in construction sites.	Water, energy
		Danish organic logo	The logo can be applied only to products on farms authorised for organic production.	Food
		Organic labelling for catering	Organic labels have been developed for marketing organic food products in large-scale kitchens, schools, and so on.	Food

(continued)

Table 8.5 (continued)

Tool	Tool type	Policy title	Description	WEF sectors addressed
	Clustering policies	Green Industrial Symbiosis	The national programme aims to promote competitiveness and resource efficiency through commercial partnerships between companies that can make use of each other's excess resources.	Water, energy, food
		Rethink Water Network	Companies, organisations, institutions, and government bodies share knowledge and collaborate on water-related issues.	Water
	Resource mapping	Online vulnerability mapping	An online tool is available for municipalities and property owners to map climate change vulnerabilities.	Water
	Public procurement	Partnership for green public procurement	The national government has joined with municipalities, regions, and a utility to create market conditions for green procurement. A Forum for Sustainable Procurement was created to serve as a knowledge hub among procurers in the public and private sector,	Water, energy, food
	Voluntary agreements	Voluntary Agreement Scheme on Energy Efficiency	A three-year contract is entered between a company, or a group of companies from the same sub-sector, and the Danish Energy Agency, with the company/companies required to implement energy management systems and undertake audits with an energy tax relief for participants.	Energy

Notes

1. Ministry of Industry, B. A. F. A. 2016. Report on growth and competitiveness–Summary.Available:http://em.dk/english/publications/2016/16-05-18-report-on-growth-and-competitiveness.
2. Ministry of Industry, B. A. F. A. 2016. Report on growth and competitiveness–Summary.Available:http://em.dk/english/publications/2016/16-05-18-report-on-growth-and-competitiveness.
3. Ministry of Foreign Affairs of Denmark. 2016. *Facts and statistics* [Online]. Available: http://denmark.dk/en/quick-facts/facts/.
4. Ministry of Foreign Affairs of Denmark. 2016. *Facts and statistics* [Online]. Available: http://denmark.dk/en/quick-facts/facts/.
5. Danish Ministry of the Environment. 2014. Water supply in Denmark. Available: http://www.geus.dk/program-areas/water/denmark/vandforsyning_artikel.pdf.
6. Ministry of Foreign Affairs of Denmark. 2017b. *A land enriched by water* [Online]. Available: http://denmark.dk/en/green-living/sustainable-projects/a-land-enriched-by-water.
7. Energinet. 2016. *Consumption in Denmark* [Online]. Available: http://www.energinet.dk/EN/KLIMA-OG-MILJOE/Miljoerapportering/Sider/Forbrug-i-Danmark.aspx.
8. Danish Energy Agency. 2014b. Energy statistics 2014. Available: https://ens.dk/sites/ens.dk/files/Statistik/energystatistics2014.pdf
9. Danish Agriculture and Food Council. 2016. Facts and figures. Denmark – a food and farming country. Available: http://www.agricultureandfood.dk/prices-statistics/annual-statistics.
10. Ibid.
11. Danish Business Authority. 2017a. *Fund for green business development* [Online]. Available: https://groenomstilling.erhvervsstyrelsen.dk/fund-green-business-development.
12. Danish Business Authority. 2017b. *Green business models* [Online]. Available: https://groenomstilling.erhvervsstyrelsen.dk/green-business-models
13. SPP Regions. 2015. Best practices in regional SPP/PPI networks: Partnership on green public procurement, Denmark. Available: http://www.sppregions.eu/fileadmin/user_upload/Resources/Denmark.pdf.
14. Danish Business Authority. 2017d. *What is green industrial symbiosis?* [Online]. Available: https://groenomstilling.erhvervsstyrelsen.dk/what-green-industrial-symbiosis

15. Danish Business Authority. 2017c. *Green industrial symbiosis* [Online]. Available: https://groenomstilling.erhvervsstyrelsen.dk/green-industrial-symbiosis.
16. Ministry of Environment and Food of Denmark. 2014. The Danish Nature Interpretation Service. Available: http://naturstyrelsen.dk/media/nst/66873/DNIS.pdf.
17. Danish Ministry of the Environment Information Centre. 2013. Protect water, nature and human health. Pesticides strategy 2013–2015. Available: http://c-ipm.org/fileadmin/c-ipm.org/Danish_NAP__in_EN_.pdf.
18. Ibid.
19. Ibid.
20. Ibid.
21. Ibid.
22. Danish Nature Agency. 2012. How to manage cloudburst and rain water. Action plan for a climate-proof Denmark. Available: http://en.klimatilpasning.dk/media/590075/action_plan.pdf.
23. Ibid.
24. Ibid.
25. Ibid.
26. Ibid.
27. State of Green. 2017. *Denmark knows water* [Online]. Available: http://www.rethinkwater.dk/about
28. Danish Energy Agency. 2013. Energy policy toolkit on energy efficiency in new buildings. Experiences from Denmark. Available: https://stateofgreen.com/files/download/395.
29. Danish Energy Agency. 2015. Energy policy toolkit on energy efficiency in industries. Experiences from Denmark. Available: https://ens.dk/sites/ens.dk/files/Globalcooperation/ee_in_industries_toolkit.pdf.
30. Ibid.
31. Ibid.
32. Ibid.
33. Ibid.
34. Danish Energy Agency. 2014a. Denmark's national energy efficiency action plan (NEEAP). Available: https://ec.europa.eu/energy/sites/ener/files/documents/2014_neeap_en_denmark.pdf.
35. Ibid.
36. Ministry of Climate, E. A. B. 2014. Strategy for energy renovation of buildings. The route to energy-efficient buildings in tomorrow's

Denmark. Available: https://ec.europa.eu/energy/sites/ener/files/documents/2014_article4_en_denmark.pdf.

37. Danish Energy Agency. 2014a. Denmark's national energy efficiency action plan (NEEAP). Available: https://ec.europa.eu/energy/sites/ener/files/documents/2014_neeap_en_denmark.pdf.

38. Ministry of Environment and Food of Denmark. 2015. Organic action plan 2015. Available: http://en.mfvm.dk/focus-on/organic-denmark/organic-action-plan-2015/.

39. Ministry of Food, A. A. F. 2013. Sustainability, efficiency and value enhancement in the Danish food industry. Available: https://naturerhverv.dk/fileadmin/user_upload/NaturErhverv/Filer/Indsatsomraader/GUDP/GUDP-bestyrelsen/Green_Development_and_Demonstration_Programme.pdf.

40. Ibid.

41. Ibid.

42. Anders Nemming and Rune Ventzel Hansen. Fertilizer accounts in Denmark. HELCOM workshop, 2015 Oldenburg. Ministry of Food, Agriculture and Fisheries of Denmark.

43. ibid.

44. Ministry of Environment and Food of Denmark. 2017a. *Financial support for applications for authorisation of alternative plant protection products* [Online]. Available: http://eng.mst.dk/topics/pesticides/grant-programmes/alternative-pesticides/.

45. Ministry of Foreign Affairs of Denmark. 2017a. *Cow dung – a source of green energy* [Online]. Available: http://denmark.dk/en/green-living/sustainable-projects/cow-dung-a-source-of-green-energy.

46. Ministry of Environment and Food of Denmark. 2017b. *Organic farming* [Online]. Available: http://agrifish.dk/agriculture/organic-farming/#c22017.

47. Ministry of Environment and Food of Denmark. 2017c. *Organic labelling for catering* [Online]. Available: http://en.mfvm.dk/focus-on/organic-denmark/organic-labelling-for-catering/.

References

Danish Agriculture and Food Council. 2016. Facts and figures. Denmark—A food and farming country. http://www.agricultureandfood.dk/prices-statistics/annual-statistics

Danish Business Authority. 2017a. *Fund for green business development* [Online]. https://groenomstilling.erhvervsstyrelsen.dk/fund-green-business-development
————. 2017b. *Green business models* [Online]. https://groenomstilling.erhvervsstyrelsen.dk/green-business-models
————. 2017c. *Green industrial symbiosis* [Online]. https://groenomstilling.erhvervsstyrelsen.dk/green-industrial-symbiosis
————. 2017d. *What is green industrial symbiosis?* [Online] https://groenomstilling.erhvervsstyrelsen.dk/what-green-industrial-symbiosis
Danish Energy Agency. 2013. Energy policy toolkit on energy efficiency in new buildings. Experiences from Denmark. https://stateofgreen.com/files/download/395
————. 2014a. Denmark's national energy efficiency action plan (NEEAP). https://ec.europa.eu/energy/sites/ener/files/documents/2014_neeap_en_denmark.pdf
————. 2014b. Energy statistics 2014. https://ens.dk/sites/ens.dk/files/Statistik/energystatistics2014.pdf
————. 2015. Energy policy toolkit on energy efficiency in industries. Experiences from Denmark. https://ens.dk/sites/ens.dk/files/Globalcooperation/ee_in_industries_toolkit.pdf
Danish Ministry of the Environment. 2014. Water supply in Denmark. http://www.geus.dk/program-areas/water/denmark/vandforsyning_artikel.pdf
Danish Ministry of the Environment Information Centre. 2013. Protect water, nature and human health. Pesticides strategy 2013–2015. http://c-ipm.org/fileadmin/c-ipm.org/Danish_NAP__in_EN_.pdf
Danish Nature Agency. 2012. How to manage cloudburst and rain water. Action plan for a climate-proof Denmark. http://en.klimatilpasning.dk/media/590075/action_plan.pdf
Energinet. 2016. *Consumption in Denmark* [Online]. http://www.energinet.dk/EN/KLIMA-OG-MILJOE/Miljoerapportering/Sider/Forbrug-i-Danmark.aspx
Ministry of Climate, Energy and Building. 2014. Strategy for energy renovation of buildings. The route to energy-efficient buildings in tomorrow's Denmark. https://ec.europa.eu/energy/sites/ener/files/documents/2014_article4_en_denmark.pdf
Ministry of Environment and Food of Denmark. 2014. The Danish Nature Interpretation Service. http://naturstyrelsen.dk/media/nst/66873/DNIS.pdf
————. 2015. Organic action plan 2015. http://en.mfvm.dk/focus-on/organic-denmark/organic-action-plan-2015/

————. 2017a. *Financial support for applications for authorisation of alternative plant protection products* [Online]. http://eng.mst.dk/topics/pesticides/grant-programmes/alternative-pesticides/

————. 2017b. *Organic farming* [Online]. http://agrifish.dk/agriculture/organic-farming/#c22017

————. 2017c. *Organic labelling for catering* [Online]. http://en.mfvm.dk/focus-on/organic-denmark/organic-labelling-for-catering/

Ministry of Food, Agriculture and Fisheries. 2013. Sustainability, efficiency and value enhancement in the Danish food industry. https://naturerhverv.dk/fileadmin/user_upload/NaturErhverv/Filer/Indsatsomraader/GUDP/GUDP-bestyrelsen/Green_Development_and_Demonstration_Programme.pdf

Ministry of Foreign Affairs of Denmark. 2016. *Facts and statistics* [Online]. http://denmark.dk/en/quick-facts/facts/

————. 2017a. *Cow dung—A source of green energy* [Online]. http://denmark.dk/en/green-living/sustainable-projects/cow-dung-a-source-of-green-energy

————. 2017b. *A land enriched by water* [Online]. http://denmark.dk/en/green-living/sustainable-projects/a-land-enriched-by-water

Ministry of Industry, Business and Financial Affairs. 2016. Report on growth andcompetitiveness—Summary.http://em.dk/english/publications/2016/16-05-18-report-on-growth-and-competitiveness

Nemming, Anders, and Rune Ventzel Hansen. 2015. Fertilizer accounts in Denmark. HELCOM workshop, 2015 Oldenburg. Ministry of Food, Agriculture and Fisheries of Denmark.

OECD. 2016. *Denmark—Economic forecast summary (November 2016)* [Online]. http://www.oecd.org/economy/denmark-economic-forecast-summary.htm

SPP Regions. 2015. Best practices in regional SPP/PPI networks: Partnership on green public procurement, Denmark. http://www.sppregions.eu/fileadmin/user_upload/Resources/Denmark.pdf

State of Green. 2017. *Denmark knows water* [Online]. http://www.rethinkwater.dk/about

Statistics Denmark. 2017. *Population projections* [Online]. http://www.dst.dk/en/Statistik/emner/befolkning-og-valg/befolkning-og-befolkningsfremskrivning/befolkningsfremskrivning

9

The Green Economy and the Water-Energy-Food Nexus in Korea

Introduction

In 1960, Korea was one of the poorest countries in the world. In 1962, the government began a series of five-year national plans for economic development, and the gross national product rose from 80 dollars in 1960 to 1600 dollars in 1980 and passed the 10,000-dollar mark in the 1990s. The economic growth resulted in Korea becoming the eighth-largest trading nation. However, over the period 2000–2010, GDP growth averaged at only 4.1 percent, a significant drop from the average of 7.9 percent over the period 1960–2000. Moreover, from 2011 to 2015 Korea's average growth was only 3 percent. Over the next ten years, it is projected that Korea's GDP growth rate will decline to around 2 percent per annum.[1] Korea has a population of 50.9 million, which is rapidly declining due to a low birth rate of only 1.19 children per mother. It is estimated that the population could decrease by 13 percent to just over 42 million in 2050.[2]

© The Author(s) 2018
R.C. Brears, *The Green Economy and the Water-Energy-Food Nexus*,
DOI 10.1057/978-1-137-58365-9_9

9.1   Water-Energy-Food Nexus Pressures

Korea is experiencing a variety of water, energy, and food nexus pressures that are detrimental to the development of a green economy as described below through a variety of examples.

Water

Korea is a country with abundant rainfall with annual average rainfall 1.4 times larger than the global average rainfall. However, annual rainfall per person is just 13 percent of the global average. In addition, two-thirds of this rainfall is concentrated in the summertime. Surface runoff accounts for 58 percent of the total water amount, of which the rest is lost as evapotranspiration. Of the total surface runoff, water from river, dam, and groundwater accounts for 10 percent, 14 percent, and 3 percent respectively, while the remaining 31 percent of surface runoff goes out to sea. This means that only 27 percent of the total water amount is usable.[3] Rapid industrialisation and economic growth have changed the pattern of water demand over time. Over the period 1975–1994, municipal and industrial water withdrawal increased steadily from 10 and 0 percent respectively to 26 and 11 percent respectively, while agricultural water withdrawal decreased from 90 to 63 percent in the same period.[4] Today, agricultural water use accounts for 48 percent of water used, making Korea one of the top 10 OECD countries in groundwater use for agricultural irrigation.[5,6]

Energy

Korea imports around 98 percent of its fossil fuel consumption due to insufficient domestic resources, making it one of the world's leading energy importers. Although petroleum and other liquids, including bio-fuels, accounted for the largest portion (41 percent) of Korea's primary energy consumption in 2015, its share has been declining since the mid-1990s, when it reached a peak of 66 percent. This trend is due to the steady increase in natural gas, coal, and nuclear energy consumption reducing oil use in the power and industrial sectors. Korea will take actions to enhance energy efficiency and increase the use of renewable energy to become more energy independent. Currently renewable sources accounted for 1 percent of Korea's primary energy consumption in 2015.[7] To reduce carbon emissions from power plants, Korea had planned to enlarge its use of nuclear energy from 26 percent in 2009 to 32 percent in 2020; however, following Japan's Fukushima disaster and false safety certifications of nuclear parts in late 2012, the government has scaled back its long-term reliance on nuclear power in the electricity portfolio that was first outlined in 2008.[8]

Food

Korea's food and beverage per capita consumption is projected to grow by 5.1 percent up to 2016, with the biggest area of growth in the fresh food market, particularly fruit, nuts, and meat. To meet this demand, area of land used for growing vegetables and fruit and for greenhouse farming which can yield more profit is increasing.[9] However, at the same time land for rice harvesting has declined steadily due to land conversions to non-agricultural uses and the country's efforts to encourage a shift towards other crops.[10] Affluent young professionals are buying non-essential food items, especially those with health and functional benefits. While consumers are generally basing their purchasing decisions on price, taste, quality, brand, and country of origin, sustainability, carbon footprint, environmental issues, and traceability are becoming important factors.[11]

9.2   General Fiscal Tools to Reduce Water-Energy-Food Nexus Pressures

Korea has implemented a variety of fiscal tools that create interdependencies and synergies between the nexus systems while reducing trade-offs between the systems in the development of a green economy.

Greening of Industry Tax Exemptions

To promote the greening of industry the government has introduced tax incentives for investing in green technology and industry. A tax exemption has been developed for dividends and interest from deposits, bonds, and investment funds that invest at least 60 percent of their total capital in firms and projects with national green certificates: Korea has established a certification system run by the Green Certification Committee that determines which technologies, projects, and firms are green based on an evaluation by the Korea Institute of Advancement of Technology. In addition, the government has expanded the loan programme for green firms and projects and launched a private equity fund to support firms, projects, and technologies that are granted green certificates.

Green Credit Scheme

Under the Green Credit Scheme major companies that assist SMEs to reduce their GHG emissions will be credited with the emissions savings. To encourage companies to use the Energy Management System more widely the government has provided sectoral application software and SME support packages. An Energy Service Companies (ESCO) loan fund and an ESCO fund have been created with the government carrying out a competitive application process to identify and support the best projects that guarantee savings.

9.3   General Non-fiscal Tools to Reduce Water-Energy-Food Nexus Pressures

Korea has implemented a variety of non-fiscal tools that create interdependencies and synergies between the nexus systems while reducing trade-offs between the systems in the development of a green economy.

GHG and Energy Target Management System

In 2010, the government introduced the GHG and Energy Target Management System for large commercial and public buildings. Under this system, the government imposes a GHG emissions target, as well as energy use limits, on designated companies with large GHG emissions and energy consumption. The government then monitors the achievements of these companies. All new buildings must meet the highest standard in the energy performance certification programme. If they exceed 10,000 square metres in area, achieve energy savings of over 5 percent, or are likely to recover the investment in ten years, they are required to propose an energy efficiency project with an energy service company.

9.4   Water: Fiscal Tools

Korea has implemented a variety of water-related fiscal tools that create interdependencies and synergies between the nexus systems while reducing trade-offs between the systems in the development of a green economy.

Water Use Charges for Source Protection

In order to protect upstream water resources that flow downstream and forge a cooperative relationship between upstream and downstream residents, Korea levies a water use charge, based on volume of water usage, on downstream residents to compensate for the land use regulations that are in effect for upstream residents, with the water use charge varying between the four major rivers (Table 9.1).[12] Revenue from the water use charge, based on volume of water usage, is then used to support upstream community projects and construct water-cleaning facilities.[13]

Water+Loan

K-Water is expanding the budget to support small businesses with R&D through a public–private joint investment fund, which involves signing a Memorandum of Understanding for the establishment of a loan programme for small businesses called 'Water+Loan' with major financial companies.[14]

Table 9.1 Water use charges in the four major rivers

Water shed/ category	Han River	Nakdong River	Geum River	Yeongsan River
Rate (won/tonne)	120	110	130	130
Total (billion won)	296.6	137.3	50.4	40
Usage	• Support upstream community projects • Purchase land in watershed areas • Support construction of upstream wastewater treatment plants			

9.5   Water: Non-fiscal Tools

Korea has implemented a variety of water-related non-fiscal tools that create interdependencies and synergies between the nexus systems while reducing trade-offs between the systems in the development of a green economy.

Smart Water Grid

Climate change impacts the availability of water, in addition to challenges of urbanisation and industrial development leading to water pollution and ecological degradation. K-Water is responding to uncertainty in water quantity and quality by developing a Smart Water Grid that combines existing water grids with information and communication technologies. The Smart Water Grid enables K-Water to have real-time monitoring of the entire water supply system to ensure adequate quantity and consistent water quality. The Smart Water Grid also comprises a sensor network inside the pipelines that collects and analyses water data including quantity, quality, pressure, and leakage. In addition, the Smart Water Grid enables customers to receive real-time information about tap water quality over the whole production and transportation process.[15]

Carbon Labelling of Tap Water

To lower carbon emissions and energy requirements in providing potable water, K-Water has begun carbon labelling tap water where carbon label-

ling certification is granted for a product that has been proved to minimise the emission of greenhouse gases in the entire cycle of its production.
In 2011, tap water produced at Cheongju waterworks qualified as being
Korea's first 'carbon labelling' tap water. By 2013, ten K-Water waterworks have been qualified for 'carbon labelling'.[16]

K-Water's One-Stop for Small Businesses

K-Water has established 'One-stop', which supports technology development for small businesses in its supply chain. One-stop provides an on-
site performance test-bed for small businesses to test their new technologies
on-site. K-Water also provides small businesses with special contracts for
their products or technology that performs well in the test-bed provided
for by K-Water. For example, K-Water has established a performance test-
bed for new technologies in hydropower in order to support the development of domestic technologies while enhancing the technological
competitiveness of small businesses.[17]

K-Water's Eco-Friendly Partnerships with Small Businesses

K-Water has established eco-friendly partnerships with small businesses
by providing various environmental management programmes including
education, technological support for environmental management, and
ISO 14001 expenses, while receiving eco-friendly products and services
from them. K-Water's customised support for small businesses leads to
various benefits including productivity improvements, increased profit
through increased customer satisfaction, and reduction in production
costs by reducing environmental costs.[18]

9.6   Energy: Fiscal Tools

Korea has implemented a variety of energy-related fiscal tools that create
interdependencies and synergies between the nexus systems while reducing
trade-offs between the systems in the development of a green economy.

Promoting High-Efficiency Appliances

To promote high-efficiency appliances—first-grade energy efficiency appliances, high-efficiency certified products, and Energy Boy–labelled products—the government uses a variety of fiscal tools including:

- *Rebates*: To promote high-efficiency appliances subsidies are provided to installation parties that install appliances and equipment approved as high-efficiency with the subsidy amount dependent on the number of appliances installed.
- *Tax breaks on energy-saving investments*: Domestic residents investing in energy-saving facilities can apply for a tax waiver (10 percent of total investment cost) from income or corporate tax.
- *Financing Rational Energy Utilization Fund*: Financial support is available for installing high-efficiency certified appliances and production facilities that manufacture high-efficiency products (Table 9.2).[19]
- *Testing fee waiver*: Korea Energy Management Corporation (KEMCO) provides financial support to small and medium-sized companies for costs incurred in testing the efficiency of appliances in the laboratory.

Table 9.2 High-efficiency appliances installation support

Type	Maximum financing per year	Loan period	Interest rate	Target recipient
Installing high-efficiency appliances	Up to 10 million USD	3 years and redemption by instalments for 5 years	Guidance on the management of energy resources project special account	Installation user
Installing high-efficiency appliances production system	Up to 5 million USD	3 years and redemption by instalments for 5 years		Manufacturers (small- and medium-sized companies)

Renewable Portfolio Standards Credit Trading

In 2012, the government introduced a Renewable Portfolio Standards (RPS) system that mandates power producers, with power generating facilities with installed capacity over 500 MW, produce a minimum proportion of their electricity from renewable energy sources. The annual RPS target rises from 2 percent in 2012 to 10 percent in 2022.[20] Under the RPS, producers of electricity from renewables receive Renewable Energy Certificates (REC) based on kilowatt-hours (kWh) of electricity produced; creating a supply of certificates. The number of certificates per kWh varies between sources based on generation costs, the expected impact on renewable technologies, and the environmental effect; for example, offshore wind production is worth two certificates per kWh compared to 0.25 for combined cycle electricity generation. Electricity companies need to purchase certificates when their direct production of renewable energy falls short, generating demand for certificates.[21] The obligatory supply can be deferred for a maximum of three years with the deferred amount a maximum 20 percent of the obligatory supply. Penalties of a maximum of 150 percent of trading prices of REC are imposed for non-compliance with the amount varying depending on causes and frequency of the non-compliance.[22]

Home Subsidy Program (1 Million Green Homes Program)

To encourage new renewable energy in homes the government provides a subsidy of 50 percent to install renewable energy systems in private residential houses, multi-family houses, and public rental houses. The programme covers a variety of resources including solar PVs, solar thermal, geothermal, small wind, and bio with the target of installing renewable energy systems in one million homes by 2020 (Table 9.3).[23]

Table 9.3 1 Million Green Homes Program

	2004–2007	2008–2012	2013–2020
Target (houses)	17,400	94,150	913,000
Accumulation (houses)	17,400	111,550	1,024,550
Investment (100 mill KRW)	2280	13,300	137,530

9.7   Energy: Non-fiscal Tools

Korea has implemented a variety of energy-related non-fiscal tools that create interdependencies and synergies between the nexus systems while reducing trade-offs between the systems in the development of a green economy.

Energy Efficiency Labelling

In Korea, energy efficiency labelling schemes play a key role in improving the energy efficiency of appliances and equipment and reducing carbon emissions. In particular, the country operates its Energy Efficiency Label and Standard Program, High-efficiency Appliance Certification Program and e-Standby Program to improve energy efficiency in appliances and equipment.

Energy Efficiency Label and Standard Program

Under this programme, manufacturers and importers are mandated to produce and sell energy-efficient products from the outset. The programme enables consumers to identify energy-efficient products easily by: first, having a mandatory energy efficiency grade from first to fifth on the product (with first-grade products saving around 30–40 percent more energy than fifth-grade products); second, mandatory reporting of energy efficiency grades by manufacturers or importers where energy efficiency labelling tests are conducted on request at designated national testing institutes (or self-certified testing institutes) with the product's test report sent to KEMCO and made publicly available online; and third, applying minimum energy performance standards that ban the production and sale of low-energy-efficiency products that fall below a minimum energy performance standard, with a fine for violations. The programme's target products include refrigerators, washing machines, and TVs.[24]

High-Efficiency Appliance Certification Program

This voluntary certification programme certifies products for industry and buildings as high-efficiency appliances if their energy efficiency and quality test results are above the certification standards set by the government. The programme's target products include LED lights and boilers. Manufacturers interested in receiving a high-efficiency appliance certification need to submit an application along with documents on maintenance of certified efficiency and a performance testing report issued by designated testing institutes.[25]

e-Standby Program

This mandatory reporting scheme encourages the adoption of energy-saving modes while the appliances are idle and the minimisation of standby power. Under this programme all e-Standby Program target products, including computers, printers, and scanners, are subject to mandatory reporting of standby power with a mandatory indication on appliances that fall below the standby power standards, with a fine for violations. Specifically, an Energy Boy label is attached to products that meet the standards for standby power, while a Standby warning label is attached to those products that do not meet the specified standby power standards, where products with an Energy Boy label save between 30 and 50 percent more energy than ordinary products.[26]

Energy Efficiency and Government Procurement

To promote high-efficiency appliances—first-grade energy efficiency appliances, high-efficiency certified products, and Energy Boy–labelled products—the government uses a variety of green procurement policies:

- *Priority purchases from the public procurement service*: When making purchases through the public procurement service, high-efficiency certified products, Energy Boy–labelled products, and first-grade energy efficiency labelled products receive priority in purchase. In addition, the public procurement service operates a high-efficiency appliance Internet shopping mall.

- *Designation of best procured products*: When making purchases through the public procurement service, high-efficiency certified products are designated as best procured products.
- *Mandatory use of energy-efficient appliances by all public organisations*: All public organisations must use high-efficiency certified appliances when making purchases or replacing existing appliances. By 2012 all public organisations had to replace more than 30 percent of their lighting equipment with LED products; select Energy Boy–labelled products when making new purchases or replacing existing appliances and install software that automatically saves power when appliances including computers are not in operation; and use first-grade energy efficiency appliances.
- *Building codes*: For newly constructed apartments, townhouses, and educational facilities with floor space of 3000 square metres or more, it is mandatory for energy-efficient transformers, lighting equipment (high-efficiency certified appliances, first-grade energy efficiency products) and automatic standby power blocking outlets or switches to be used, and recommended that other appliances including domestic gas boilers be first-grade efficient.
- *Standards and performance of eco-friendly houses*: For residential housing complexes with more than 20 families, mandatory regulations apply to ensure they are eco-friendly; for example, fluorescent lamps must be first-grade and automatic standby power blocking outlets and switches must be installed.[27]

Smart Grid Demonstration and Testing

Korea's roadmap for a national smart grid aims to complete construction of a smart grid across metropolitan areas and the fitting of all households with smart meters by 2020. The goal is to complete a nationwide smart grid by 2030. The development of a smart grid, along with an increase in renewable energy by this date, will lead to significant environmental and economic benefits: elimination of approximately 230 million tonnes of GHG emissions, the creation of 50,000 jobs annually, and removal of the need to construct USD 2.8 billion worth of new power plants.

Testing the Smart Grid

To test the concept of smart grids, the Korean Ministry of Knowledge Economy established the Jeju Island Smart Grid Test-Bed project. The project, involving 168 companies, aims to develop a smart grid that will collect real-time data on energy usage and demand with the data used to limit the unnecessary use of electricity and increase the efficiency of energy consumption. The first phase of the project involved constructing smart grid facilities and infrastructure and the second phase testing and demonstrating the infrastructure and services, before preparing the technology for expansion in mainland Korea.

One of the key aspects of the smart grid will be enabling households and businesses to move from being passive consumers of electricity to well-informed and proactive users. The project involves testing a wide range of connected devices, sensors, and web-based cloud services to analyse and present energy consumption data, including installation of smart tags that monitor the power usage of each electrical device and the connection of household appliances to smart boxes enabling the control of household appliances via wide area networks (WAN). This will be combined with temperature and humidity sensors to allow real-time monitoring and control of energy consumption. The project will also test the development of an energy consumption portal in which households can view their energy consumption data on four different screens including TVs, tablets, and smartphones. Furthermore, every household will be able to compare its usage with comparable houses and neighbours.[28]

Audit System

The Energy Audit System, introduced in 2007 under the Energy Use Rationalisation Act, aims to conserve energy and reduce GHG emissions in the industrial and building sectors. This is done by regularly inspecting businesses that consume more than 2000 tonnes of oil equivalent (TOE) per year. For businesses that consume less than 10,000 TOE per year the cost of the audit will be subsidised by up to 90 percent.

Eco-Friendly Home Codes

In 2009, the government implemented eco-friendly home codes to ensure low-energy-consuming homes. Blocks of flats with over 20 households must be designed to be 20 percent more energy efficient than existing buildings. In 2010, the government provided KRW 14 billion for retrofitting homes and mandated that buildings containing over 300 households provide their energy consumption on an Internet portal.

Mandatory Use of Renewable Energy in Buildings

Since 2011, energy supply for new buildings and extended or reconstructed buildings that exceed 3000 square metres must include at least 10 percent new and renewable energy. The obligation will be gradually increased to 20 percent in 2020.[29]

9.8   Food: Fiscal Tools

Korea has implemented a variety of food-related fiscal tools that create interdependencies and synergies between the nexus systems while reducing trade-offs between the systems in the development of a green economy.

Pay-as-You-Throw

The Ministry of Environment has set a nationwide 40 percent food waste reduction target. To achieve the target the government in 2005 implemented a 'pay-as-you-throw' programme in selected areas on a trial basis. By 2012, 95 percent of the country's 153 self-governing municipalities had implemented the system. Under the programme the Ministry of Environment accepts three billing methods for municipalities to charge residents with each municipality deciding on which method to use, with some accepting multiple methods. The three methods are:

- *Radio frequency identification (RFID) food waste management system*: When users tap a card with their personal RFID tag over a reader on a specially designed food waste recycling bin, the bin's lid will automatically open. Waste discarded into the bin is then automatically weighed and recorded under the user's account. The user is then billed monthly based on the total weight of dumped food waste in that particular month.
- *Prepaid garbage bags*: Residents pre-pay for waste collection when purchasing specially designed garbage bags which are priced based on volume.
- *Bar code management system*: Consumers deposit food waste into composting bins of standardised sizes and pay disposal fees by purchasing either bar code stickers attached to the bin or bar codes inserted into a reader on the bin.[30]

9.9   Food: Non-fiscal Tools

Korea has implemented a variety of food-related non-fiscal tools that create interdependencies and synergies between the nexus systems while reducing trade-offs between the systems in the development of a green economy.

Agriproduct Certification

The Korean government enforces a certification system that ensures the quality of Korean agricultural products. To help consumers in Korea and abroad identify high-quality and reliable agriproducts Korea AgraFood has introduced a range of agriproduct-related certifications:

Certification of Quality for Traditional Food Products

The traditional food quality certification system guarantees the quality of the traditional Korean foods whose taste, scent, and colour are inherent to Korea, which are made with homegrown food ingredients and processed using age-old traditional methods. The Agriculture, Food and Rural Affairs Ministry determines which items can obtain this certification after an examination of various factors including marketability, appeal, and traditionalism.

Certification for Organically Processed Food

The organically processed food certification system was developed to improve the reliability of organic agriculture, to protect consumers, and to encourage agriproduct producers and suppliers to provide organically processed high-quality food, where organically processed food refers to produce and livestock.

Good Agricultural Practices Certification

Good agricultural practices stand for the standards to control hazardous factors including agricultural chemicals, heavy metals, or harmful organisms that can remain in the agricultural environment, including soil and water, or in agricultural products. To receive certification, standards should be applied throughout the entire process from cultivation and harvesting to packaging in order to secure the food safety of products.

Protecting Geographical Indication System

The protecting geographical indication system promotes the place-of-origin indication in agriproducts whose quality, reputation, or other characteristics originate from a certain geographical region or area.[31]

9.10 Case Study Summary

Korea has implemented a variety of initiatives to create interdependencies and synergies between the nexus systems while reducing trade-offs between the systems in the development of a green economy.

General

- Korea offers tax exemptions for investments in green technology and industries that are eligible to receive green certificates.
- Under the Green Credit Scheme, large companies that assist SMEs in reducing emissions can be credited with the emission savings.
- Korea imposes GHG emission targets and energy use limits on designated companies. In addition, all new buildings must meet a high energy-saving standard with large buildings or buildings that can make significant energy savings required to submit energy efficiency projects.

Water

- Korea levies water charges on downstream water users of the four major rivers with the revenue collected used to support upstream community projects and construct water-cleaning facilities.
- K-Water provides loans to small businesses undertaking water technology–related R&D projects.
- K-Water has set up a demonstration smart water grid that provides real-time water quantity and water quality data.

- K-Water has begun carbon labelling of tap water from various water works in an attempt to reduce GHG emissions and energy requirements in providing potable water.
- K-Water has established a one-stop shop for small businesses to test their technology at K-Water sites, from which contracts are given to small businesses whose technologies perform well in the test-bedding phase.
- K-Water has also established eco-friendly partnerships with small businesses in which K-Water provides various environmental management programmes while receiving eco-friendly products and services.

Energy

- Korea uses a variety of fiscal instruments including rebates and tax breaks on energy-saving appliances. The government also provides financial support for installing high-efficiency certified appliances and production facility systems.
- Power producers must produce a minimum proportion of their electricity from renewable energy sources. If producers fall short of their requirements they can purchase certificates from other producers, creating a market for these certificates.
- Korea provides a subsidy to install renewable energy systems in residential houses and complexes with the subsidy covering a variety of renewable resources including solar- and wind-power as well as bioenergy.
- Korea mandates manufacturers and importers to produce and sell energy-efficient products. In addition, the country has an e-Standby programme that mandates that appliances have a standby label that signifies the product meets standby power standards or a standby warning label if the product fails to meet these standards.
- Korea also has a voluntary certification programme which certifies products for industry and buildings as high-efficiency appliances if their energy efficiency and quality tests are above the certification standards set by the government.

- Korea's government has a variety of green procurement policies to ensure high-efficiency certified products are purchased including placing energy-efficient items on priority purchase lists, operating a high-efficiency appliance Internet shopping mall, mandating use of energy-efficient appliances in all public organisations, and mandating large-scale new buildings and new residential housing complexes install high-efficiency products.
- Korea is developing a Smart Grid test-bed that involves the government along with companies developing infrastructure and data systems to enable the real-time collection of data on energy usage and demand.
- All new buildings or reconstructed buildings that exceed a certain size must include at least 10 percent new and renewable energy in the designs, increasing to 20 percent in 2020.

Food

- Korea has implemented a pay-as-you-throw food waste reduction programme in selected areas on a trial basis. The ministry in charge enables municipalities in the test areas to choose three billing methods to charge customers for collecting food waste, including tapping an RFID tag over a food recycling bin lid, prepaid garbage bags for food waste, and purchasing bar code stickers.
- Korea has an agricultural certification scheme which certifies food as being: traditional; organically processed; grown from farms that use good agricultural practices in protecting soil and water quality; or from a geographical location.

Overall, Korea uses a variety of fiscal and non-fiscal tools to create interdependencies and synergies between the nexus systems while reducing trade-offs between the systems in the development of a green economy. These tools are summarised in Table 9.4.

Table 9.4 Korea case summary

Tool	Tool type	Policy title	Description	WEF sectors addressed
Fiscal	Market-based instruments and pricing	Water use charge	The water use charge for downstream water users is used to support upstream water protection projects.	Water
		Water+Loan	Loans for small businesses to develop water technologies will be made through a public–private joint investment fund.	Water
		Green credit scheme	Major companies that assist SMEs in reducing GHG emissions are credited with the emission savings.	Energy
		Financial Rational Energy Utilization Fund	Financial support available for installation or production of high-efficiency certified appliances.	Energy
		Test fee waiver	SMEs are provided test-fee waivers for testing equipment in KEMCO labs.	Energy
		Renewable Portfolio Standards certificate trading	Electricity companies must produce a minimum amount of renewable energy. If a shortfall occurs, they can purchase certificates from other electricity companies.	Energy
		Pay-as-you-throw	Residents are charged for throwing away food to reduce food waste.	Food
	Environmental taxes	Greening of industry tax exemptions	Tax exemption for dividends and interest that originate from investments in green technology and industry.	Water, energy, food

(continued)

Table 9.4 (continued)

Tool	Tool type	Policy title	Description	WEF sectors addressed
	Financial incentives	Energy efficiency tax breaks	Domestic residents investing in energy-saving facilities can apply for a tax waiver.	Energy
		Energy efficiency rebate	Rebates are provided for the installation of high-efficiency appliances.	Energy
		Home Subsidy Program	A subsidy is provided to install renewable energy systems in houses.	Energy
Non-fiscal	Regulations	GHG and Energy Target Management System	GHG emission targets and energy use limits are imposed on designated companies with large GHG emissions and energy consumption.	Energy
		Energy Audit System	Inspections of industrial and building sector companies that consume a minimum amount of energy.	Energy
		Mandatory renewable energy in buildings	All new or renovated buildings must include a minimum amount of renewable energy.	Energy
	Standards and mandatory labelling	Energy efficiency label	Manufacturers and importers are mandated to produce and sell energy-efficient products, identifiable from energy efficiency labels.	Energy
		e-Standby	Energy Boy labels are attached to products that meet standards for standby power.	Energy

(continued)

Table 9.4 (continued)

Tool	Tool type	Policy title	Description	WEF sectors addressed
		Eco-friendly home codes	The codes ensure low-consuming homes in the future and large-scale housing provide their energy consumption on an Internet portal.	Energy
	Public–private partnerships	K-Water's eco-friendly partnerships	Eco-friendly partnerships established with small businesses to provide environmental management knowledge while receiving eco-friendly products from them.	Water, energy
	Voluntary labelling	Carbon labelling of water	K-Water provides carbon labelling for tap water produced with minimal amounts of GHG emissions.	Water, energy
		High-efficiency appliance certification	Products can be certified as high-efficiency if they exceed government standards.	Energy
		Agriproduct certification	Agriproduct certificates ensure consumers can identify high-quality and reliable Korean food.	Food, water
	Public procurement	Energy efficiency and government procurement	The government uses a variety of green procurement policies to promote energy efficiency appliances, high-efficiency certified products and Energy Boy label products.	Energy

(continued)

Table 9.4 (continued)

Tool	Tool type	Policy title	Description	WEF sectors addressed
	Demonstration projects	Smart water grid	The Smart water grid enables real-time monitoring of both water quantity and quality.	Water
		Smart Grid demonstration and testing	The government has established a test-bed project that involves companies developing and testing smart grid infrastructure and technologies.	Energy
		K-Water One-stop shop	K-Water provides an on-site performance test-bed for small businesses testing their new water technologies, with well-performing technologies receiving special contracts for their products.	Water

Notes

1. Lee, J.-W. 2016. The Republic of Korea's economic growth and catch-up: Implications for the People's Republic of China. *ADBI Working Paper Series.*
2. World Population Review. 2017. *South Korean population 2017* [Online]. Available: http://worldpopulationreview.com/countries/south-korea-population/.
3. OECD. 2015b. Water resources allocation: Korea.
4. FAO. 2011. *Republic of Korea* [Online]. Available: http://www.fao.org/nr/water/aquastat/countries_regions/KOR/.
5. OECD. 2015b. Water resources allocation: Korea.
6. OECD. 2015a. Policies to manage agricultural groundwater use: Korea. Available: https://www.oecd.org/tad/sustainable-agriculture/groundwater-country-note-KOR-2015%20final.pdf.
7. U.S. Department of Energy. 2017. *Korea, South* [Online]. Available: https://www.eia.gov/beta/international/analysis.cfm?iso=KOR.
8. Ibid.
9. New Zealand Trade and Enterprise. 2013. *Food and beverage market in South Korea* [Online]. Available: https://www.nzte.govt.nz/en/export/market-research/food-and-beverage-market-in-south-korea/.
10. FAO. 2016. *Country briefs: Republic of Korea* [Online]. Available: http://www.fao.org/giews/countrybrief/country.jsp?code=KOR.
11. New Zealand Trade and Enterprise. 2013. *Food and beverage market in South Korea* [Online]. Available: https://www.nzte.govt.nz/en/export/market-research/food-and-beverage-market-in-south-korea/.
12. Ibid.
13. Lee, Y.-K. 2004. Two-page summaries of innovative practices. *UNEP 8th GCSS/GMEF.* Jeju, Korea.
14. K-Water. 2014. 2014 K-water sustainability report. Available: https://www.unglobalcompact.org/participation/report/cop/create-and-submit/detail/103051.
15. Ibid.
16. Ibid.
17. Ibid.
18. Ibid.
19. KEMCO. 2011. Korea's energy standards and labeling. Available: http://www.kemco.or.kr/nd_file/kemco_eng/KoreaEnergyStandards&Labeling.pdf.

20. Korea Energy Agency. 2017. *Renewable portfolio standards (RPS)* [Online]. Available: http://www.energy.or.kr/renew_eng/new/standards. aspx.

21. Yoo, R. S. J. A. B. 2012. Achieving the "low carbon, green growth" Vision in Korea. Available: http://www.oecd-ilibrary.org/economics/achieving-the-low-carbon-green-growth-vision-in-korea_5k97gkdc52jl-en?crawler=true.

22. Korea Energy Agency. 2017. *Renewable portfolio standards (RPS)* [Online]. Available: http://www.energy.or.kr/renew_eng/new/standards. aspx.

23. KEMCO. 2017. *1 million green homes program* [Online]. Available: http://www.kemco.or.kr/new_eng/pg02/pg02040602.asp.

24. KEMCO. 2011. Korea's energy standards and labeling. Available: http://www.kemco.or.kr/nd_file/kemco_eng/KoreaEnergyStandards& Labeling.pdf.

25. Ibid.

26. Ibid.

27. Ibid.

28. GSMA. 2012. South Korea: Developing next generation utility networks in Jeju island. Available: http://www.gsma.com/connectedliving/south-korea-developing-next-generation-utility-networks-in-jeju-island/.

29. Canada Korea Business Council. 2016. Product market study. Renewable energy market in Korea. Available: http://ckbc.ca/wp-content/uploads/2016/01/Renewable-Energy-Market-in-Korea.pdf.

30. Asia Today. 2013. *South Korea's food waste solution: You waste, you pay* [Online]. Available: http://www.asiatoday.com/pressrelease/south-koreas-food-waste-solution-you-waste-you-pay.

31. AGRAFOOD. 2013. *High quality ensured by Agriproduct certification!* [Online]. Available: http://www.kfoodstory.com/news/articleView. html?idxno=3038.

References

AgraFood. 2013. *High quality ensured by agriproduct certification!* [Online]. http://www.kfoodstory.com/news/articleView.html?idxno=3038

Asia Today. 2013. *South Korea's food waste solution: You waste, you pay* [Online]. http://www.asiatoday.com/pressrelease/south-koreas-food-waste-solution-you-waste-you-pay

Canada Korea Business Council. 2016. Product market study. Renewable energy market in Korea.http://ckbc.ca/wp-content/uploads/2016/01/Renewable-Energy-Market-in-Korea.pdf

FAO. 2011. *Republic of Korea* [Online]. http://www.fao.org/nr/water/aquastat/countries_regions/KOR/

———. 2016. *Country briefs: Republic of Korea* [Online]. http://www.fao.org/giews/countrybrief/country.jsp?code=KOR

GSMA. 2012. South Korea: Developing next generation utility networks in Jeju Island. http://www.gsma.com/connectedliving/south-korea-developing-next-generation-utility-networks-in-jeju-island/

KEMCO. 2011. Korea's energy standards and labeling. http://www.kemco.or.kr/nd_file/kemco_eng/KoreaEnergyStandards&Labeling.pdf

———. 2017. *1 million green homes program* [Online]. http://www.kemco.or.kr/new_eng/pg02/pg02040602.asp

Korea Energy Agency. 2017. *Renewable portfolio standards (RPS)* [Online]. http://www.energy.or.kr/renew_eng/new/standards.aspx

K-Water. 2014. 2014 K-water sustainability report. https://www.unglobalcompact.org/participation/report/cop/create-and-submit/detail/103051

Lee, Y.-K. 2004. Two-page summaries of innovative practices. *UNEP 8th GCSS/GMEF.* Jeju, Korea.

Lee, J.-W. 2016. *The Republic of Korea's economic growth and catch-up: Implications for the People's Republic of China.* ADBI Working Paper Series.

New Zealand Trade and Enterprise. 2013. *Food and beverage market in South Korea* [Online]. https://www.nzte.govt.nz/en/export/market-research/food-and-beverage-market-in-south-korea/

OECD. 2015a. Policies to manage agricultural groundwater use: Korea. https://www.oecd.org/tad/sustainable-agriculture/groundwater-country-note-KOR-2015%20final.pdf

———. 2015b. Water resources allocation: Korea.

U.S. Department of Energy. 2017. *Korea, South* [Online]. https://www.eia.gov/beta/international/analysis.cfm?iso=KOR

World Population Review. 2017. *South Korean population 2017* [Online]. http://worldpopulationreview.com/countries/south-korea-population/

Yoo, B., and R. S. Jones. 2012. Achieving the "low carbon, green growth" vision in Korea. http://www.oecd-ilibrary.org/economics/achieving-the-low-carbon-green-growth-vision-in-korea_5k97gkdc52jl-en?crawler=true

10

The Green Economy and the Water-Energy-Food Nexus in the Colorado River Basin

Introduction

The Colorado River is the single most important source of water in the southwestern United States providing water and power for nearly 40 million people and water to irrigate more than 5 million acres of farmland across seven states and 22 Native American tribes. The river also provides water for two dozen national parks, wildlife refuges, and recreation areas.[1] The Colorado River Basin covers 246,000 square miles, including parts of the seven 'basin' states of Arizona, California, Colorado, Nevada, New Mexico, Utah, and Wyoming, and also flows into Mexico.[2] Water from the Colorado River Basin is apportioned to users in the seven basin states and adjacent areas that receive water according to a series of federal laws and agreements beginning with the Colorado River Compact of 1922, which initially allocated 15 million acre-feet (maf) of water equally among Colorado, New Mexico, Utah, and Wyoming (known as the Upper Basin States) and Arizona, California, and Nevada (known as the Lower Basin States). Each basin is entitled to use of 7.5 maf per year.[3]

© The Author(s) 2018
R.C. Brears, *The Green Economy and the Water-Energy-Food Nexus*,
DOI 10.1057/978-1-137-58365-9_10

10.1   Water-Energy-Food Nexus Pressures

The Colorado River Basin is experiencing a variety of water, energy, and food nexus pressures that are detrimental to the development of a green economy as described below through a variety of examples.

Water

Demand for Colorado River water in the Lower Basin exceeds its 7.5 maf water allocation in the Compact, and while Upper Basin States have never used their full entitlement of 7.5 maf demand for water continues to grow in the Upper Basin States.[4] As such, the Colorado River Basin is likely to see large supply–demand imbalances greater than 3.5 maf for the next 50 years due to a variety of factors including climate change, drought, population growth, and environmental needs.[5] The population in the Southwest is expected to increase by nearly 70 percent by mid-century;[6] for example, Arizona's population is projected to more than double in the next 50 years with the population of Phoenix projected to grow significantly from 3.9 million in 2015 to 5.9 million in 2040.[7] The river basin will be impacted by climate change with projections indicating that by 2075 basin-wide temperatures will increase between 2.7 and 4.4 °C relative to historical averages and that basin-wide runoff is predicted to decrease by 8–10 percent relative to historical levels, with this decrease accompanied by an 8–11 percent decrease in streamflow and a 10–13 percent decrease in reservoir storage.[8]

Energy

The Colorado River Basin provides 53 gigawatts of power generation capacity;[9] three of the basin states—California, Colorado, and New Mexico—are among the United States' top 10 energy-producing states. In 2009, 87 percent of total consumption was met with fossil fuels.[10] The river basin has immense capacity to generate hydropower with hundreds of hydroelectric dams along the river's main stem and tributaries generating over 4000 MW. Over 80 percent of the river's hydropower capacity comes from the Hoover and Glen Canyon Dams. The Hoover Dam has a generation capacity of over 2000 MW and generates about 4000 GWh of electricity for customers in Nevada, Arizona, and California; the largest portion of this electricity goes to the Metropolitan Water District of Southern California (about 30 percent). Meanwhile, the Glen Canyon Dam has a capacity of over 1300 MW.[11] Climate change is expected to increase peak period electricity demands for cooling in the Southwest. In California, peak energy demands are projected to increase between 10 and 20 percent by the end of the century due to increased summer afternoon temperatures.[12]

Food

Agriculture is a major component of the river basin's economy with the basin states producing around $37 billion in crops and $24 billion in livestock in 2012. Overall, the seven states produce approximately 15 percent of all crops in the United States and around 13 percent of all US livestock. Much of the river basin is extremely arid with some areas receiving less than three inches of precipitation each year. Therefore, irrigation and agriculture are closely linked in the basin with more than 90 percent of pasture and cropland in the basin receiving supplemental water to make the land viable for agriculture. This irrigated land extends across 3.2 million acres within the basin, while water exported from the basin helps irrigate another 2.5 million acres in Colorado, Utah, New Mexico, and southern California. Overall, this type of irrigation consumes around 70 percent of the basin's water supply, not including evaporation or exports. Of the land in the basin, 60 percent is planted in forage and pasture. Alfalfa, a type of forage crop used to feed livestock including dairy cows, occupies more than a quarter of all irrigated land in the basin.[13] Climate change impacts agricultural production in the river basin with studies indicating that a 1.5 °C increase in temperature accompanied by a 10 percent decrease in precipitation could reduce agricultural welfare by 32.2 percent, and a 2.5 °C increase along with a 10 percent decrease could result in a 41.3 percent decrease respectively.[14]

10.2   General Fiscal Tools to Reduce Water-Energy-Food Nexus Pressures

Cities and states in the Colorado River Basin have implemented a variety of fiscal tools that create interdependencies and synergies between the nexus systems while reducing trade-offs between the systems in the development of a green economy.

San Diego's Green Building Incentive Program

San Diego County's Green Building Program is concerned with protecting the environment and encouraging homeowners to build using environmentally sound practices. Sustainable building practices go beyond energy and water conservation to incorporate environmentally sensitive site planning, resource-efficient building materials, and superior indoor environmental quality. Some of the key benefits include:

• Lower energy and water utility bills
• Environmentally effective use of building materials
• Enhanced health and productivity
• Long-term economic returns
• Reduced environmental impact

Green Building Incentive Program

The Green Building Incentive Program is designed to promote the use of resource-efficient construction materials, water conservation, and energy efficiency in new and renovated residential and commercial buildings. The programme offers a reduced plan check turn-around time and a 7.5 percent reduction in plan check and building permit fees for projects meeting programme requirements.[15] To qualify, the project must comply with one of the resource conservation measures listed in Table 10.1.[16]

Table 10.1 Green Building Incentive Program conservation measures

Measure	Type	Qualifying action
Natural resource conservation	Straw bale construction	New buildings using baled straw from harvested grain for the construction of exterior walls.
	Recycled content	A builder would be eligible by showing that either 20 percent or more of the primary building materials used contain a minimum of 20 percent recycled content materials, or at least one primary building material contains 50 percent or more recycled content.
Water conservation	Grey water systems	Install grey water system in new or renovated buildings.
Energy conservation	Energy use below California Energy Commission standards	Residential projects that exceed standards by 15 percent and commercial projects that exceed the standards by 25 percent.

10.3   Water: Fiscal Tools

Cities and states in the Colorado River Basin have implemented a variety of water-related fiscal tools that create interdependencies and synergies between the nexus systems while reducing trade-offs between the systems in the development of a green economy.

Residential Grey Water Subsidies and Rebate

The San Francisco Public Utilities Commission (SFPUC) is partnering with the Urban Farmer Store to offer residents a $125 discount off the purchase of a laundry-to-landscape (L2L) grey water kit that normally retails for $175. The grey water kit includes the basic components necessary to divert laundry water to the garden. In addition to the rebate, customers will receive a copy of the San Francisco Graywater Design Manual for Outdoor Irrigation; an optional on-site consultation with a grey water expert to answer questions; mandatory training to review the design, installation, and maintenance requirements of a L2L grey water system; and access to a free tool kit with the tools necessary for DIY installation. For grey water projects that require a permit from the Department of Building Inspection, required for all other grey water systems except L2L, SFPUC offers a rebate up to $225 to help cover the costs of obtaining a permit from the city's Department of Building Inspection for the installation of a grey water system in a residential unit for subsurface outdoor irrigation.[17]

Los Angeles's Water Technical Assistance Program

The Los Angeles Department of Water and Power's (LADWP) Technical Assistance Program is a financial incentive programme that offers commercial, industrial, institutional, and multi-family residential customers in Los Angeles up to $250,000 for the installation of pre-approved equipment and products that demonstrate water savings. The benefits of the programme are that customers can modernise their facility with the latest equipment, save on their water bills, and conserve water. The incentive amount received is based on the water savings accomplished by the project with the incentive calculated at $1.75 per 1000 gallons of water saved over a number of years, depending on the project, not to exceed the installed cost of the project.[18]

Tucson's Audit Program

The City of Tucson's Water Department's Drought preparedness and response plan includes four drought response levels beginning with Stage 1 and increasing in severity to Stage 4. The drought plan states that if Stage 2 drought is declared, all commercial and industrial customers, using on average over 325 ccf per month, need to conduct a self-audit of water use at their facility and develop a conservation plan. To be prepared for a Stage 2 drought, Tucson Water's Tucson Audit Program (TAP) is offering TAP audits free of charge now, to help customers get ahead of these requirements. While the programme is for the city's largest water users, any commercial or industrial customer can receive a free water audit.[19]

TAP Rebates

Tucson Water offers the TAP rebate to provide Tucson-area businesses with an incentive to act on the recommendations identified in the TAP audit's water efficiency recommendations report with the rebate offer based on the potential water savings identified in the report and existing rebates available. The rebate values are calculated as follows:

* For high-efficiency toilet and urinal replacements, the rebate is $75 per tank toilet, $150 per flush valve toilet, and $200 per urinal.
* Tucson Water provides low-flow, WaterSense-approved showerheads, low-flow aerators, and pre-rinse kitchen sprayers at no cost.
* For the remaining customised items, the rebates are calculated at $10 per 1000 gallons of projected annual savings, or 50 percent of the item cost not including labour, whichever is less.[20]

California's Water-Energy Grant Program

California's Water-Energy Grant Program provides funds to implement water efficiency programmes or projects that reduce GHG emissions and reduce water and energy use. With funds of $19 million, local agencies, joint power authorities, and non-profit organisations can seek funding of up to $3 million per proposal for programmes/projects that focus on:

* Commercial water efficiency or institutional water efficiency programmes
* Residential water efficiency programmes that benefit disadvantaged communities
* Projects that reduce GHG emissions, reduce water and energy use
* Only projects with water conservation measures that also save energy[21,22]

San Francisco's Pilot Community Garden Irrigation Meter Grant Program

To help urban agriculture become more efficient in production—more crop per drop—SFPUC encourages applications for its Pilot Community Garden Irrigation Meter Grant Program to enable urban farms, as well as community and demonstration gardens, track and manage irrigation water use through the installation of a dedicated irrigation water service. With a maximum grant of $10,000, the programme enables the installation of a dedicated irrigation service and meter to track irrigation water use as well as enable users to separately shut off the irrigation system if needed. To be eligible for SFPUC's grant proposed activities must meet one of the following criteria:

• An urban agricultural project with proof of receiving, or intent to receive, a change-of-use permit from the San Francisco Planning Department under its Urban Agricultural Ordinance approving an urban agricultural site
• A small-scale urban market garden designed and operated for urban food production
• A community garden sponsored by the San Francisco Recreation and Parks Department with a signed Community Garden Plot Agreement
• A garden established for demonstration or other educational purposes[23]

Southern Nevada's Water Efficient Technologies Programme

Southern Nevada Water Authority's (SNWA) Water Efficient Technologies (WET) programme provides financial incentives for capital expenditures when businesses retrofit existing equipment with more water-efficient consumptive-use technologies (mainly outdoor innovations that conserve water that is not returned to the sanitary sewer) and non-consumptive-use technologies (mainly innovations that conserve water returned to the sanitary sewer). Specifically:

- *Consumptive-use technologies*: Businesses can earn up to 50 percent of the product purchase price (excluding labour and installation) or $25 per 1000 gallons of water saved annually
- *Non-consumptive-use technologies*: Businesses can earn up to 50 percent of the product purchase price (excluding labour and installation) or $8 per 1000 gallons saved annually

As part of the programme, custom technology requirements include:

- Achieving annual water savings of at least 100,000 gallons
- Ensuring the rebate projects are maintained for at least ten years
- Enabling businesses to pool savings across multiple locations to reach the 100,000-gallon threshold
- Choosing SNWA-approved, standard technologies (including high-efficiency toilets, efficient shower head retrofits, waterless and high-efficiency urinal retrofits, converting sports fields from grass to artificial surfaces, retrofitting standard cooling towers with more efficient technologies)[24]

California's State Water Efficiency Enhancement Program

California's Department of Food and Agriculture's State Water Efficiency Enhancement Program (SWEEP) provides financial assistance in the form of grants to implement irrigation systems that reduce GHG emissions and save water on California agricultural operations. With $18 million in funds, SWEEP provides maximum grant awards of $200,000 with a recommended 50 percent match of the total project cost. The maximum grant duration is 12 months with projects recommended to incorporate several project types to achieve both GHG emission reductions and water savings including:

- Installing weather-, soil-, or plant-based sensors for irrigation
- Installing micro-irrigation or drip systems
- Improving fuel efficiency of water pumps

- Fuel conversion of water pumps
- Reducing pumping[25]

10.4   Energy: Fiscal Tools

Cities and states in the Colorado River Basin have implemented a variety of energy-related fiscal tools that create interdependencies and synergies between the nexus systems while reducing trade-offs between the systems in the development of a green economy.

Los Angeles' Custom Performance Program

The LADWP's Custom Performance Program (CPP) offers rebates for commercial customers to install various energy efficiency measures. LADWP customers can receive a one-time payment for investments in eligible projects that reduce electricity use in their building or facility. Most CPP incentive payments are based on the annual kilowatt hour (kWh) savings as calculated or accepted by the LADWP (examples are included in Table 10.2).[26]

Nonprofit Energy Efficiency Program

Denver's Department of Strategic Partnership's Nonprofit Energy Efficiency Program provides upgrades to non-profit facilities that serve low-income individuals and families, reducing their energy costs,

Table 10.2   Examples of Custom Performance Program incentives

Incentive categories	Incentive levels
Controls	$0.15 per kWh
Plugs	$0.15 per kWh
HVAC-Refrigeration	$0.25 per kWh
Envelope	$0.25 per kWh
LED lighting	$0.15 per kWh

ensuring more of their funding can be allocated towards programming. The programme fully funds the cost of the audit and provides upgrades with no financial contributions from the non-profit. Participating facilities have received funding between $10,000 and $120,000 for their facilities, depending on the size of their facility.[27]

Nevada's Green Building Tax Abatement

The Nevada Governor's Office of Energy (GOE) administers the green building tax abatement programme in which building owners may be eligible for a property tax abatement for constructing new buildings or renovating existing buildings (EB) to LEED or Green Globes standard; specifically, buildings must meet the equivalent of silver under the LEED standard or rating of two globes under the Green Globes standard. For new buildings, the abatement period is ten years while EB are limited to five years and capped at $100,000 per year. The specific tax abatement for new construction (NC) and core and shell (CS) as well as EB are summarised in Tables 10.3 and 10.4.[28,29,30]

Charge Up LA!: Electric Vehicle Charger Commercial Rebate Programme

The LADWP has relaunched its commercial EV charger rebate programme to expand the city's EV charging infrastructure for the home, workplace, and for stops in between. The programme offers incentives up to $4000 to LADWP commercial customers, including multi-family building owners, who purchase and install a qualified workplace or public Level 2 EV charger (240-volt). Additional connectors, beyond the first one on the same charger, are eligible for an incentive of up to $750 per connector. The rebate is available to commercial customers who have a minimum of three parking spaces available to employees, customers, visitors, or tenants. One additional Level 2 charger rebate will be available for every ten additional parking spaces at the same location, business, or

Table 10.3 Green building tax abatement for new buildings and core and shell projects

	5–6 points in the Optimize Energy Performance credit category (LEED) or 32–39 points in the Energy Performance section (Green Globes)	7–10 points in the Optimize Energy Performance credit category (LEED) or 40–55 points in the Energy Performance section (Green Globes)	11–12 points in the Optimize Energy Performance credit category (LEED) or 56–63 points in the Energy Performance section (Green Globes)	13–14 points in the Optimize Energy Performance credit category (LEED) or 64–71 points in the Energy Performance section (Green Globes)	15–16 points in the Optimize Energy Performance credit category (LEED) or 72–79 points in the Energy Performance section (Green Globes)	17–21 points in the Optimize Energy Performance credit category (LEED) or 80–100 points in the Energy Performance section (Green Globes)
LEED or Green Globes certification level						
Silver level or 2 globes	25 percent abatement for 5 years	25 percent abatement for 6 years	25 percent abatement for 7 years	25 percent abatement for 8 years	25 percent abatement for 9 years	25 percent abatement for 10 years
Gold level or 3 globes	25 percent abatement for 5 years	30 percent abatement for 6 years	30 percent abatement for 7 years	30 percent abatement for 8 years	30 percent abatement for 9 years	30 percent abatement for 10 years
Platinum level or 4 globes	25 percent abatement for 5 years	30 percent abatement for 6 years	35 percent abatement for 7 years	35 percent abatement for 8 years	35 percent abatement for 9 years	35 percent abatement for 10 years

Table 10.4 Green building tax abatement for existing buildings

LEED or Green Globes certification level	5–6 points in the Optimize Energy Performance credit category (LEED) or 32–39 points in the Energy Performance section (Green Globes)	7–10 points in the Optimize Energy Performance credit category (LEED) or 40–55 points in the Energy Performance section (Green Globes)	11–12 points in the Optimize Energy Performance credit category (LEED) or 56–63 points in the Energy Performance section (Green Globes)	13–14 points in the Optimize Energy Performance credit category (LEED) or 64–71 points in the Energy Performance section (Green Globes)	15–16 points in the Optimize Energy Performance credit category (LEED) or 72–79 points in the Energy Performance section (Green Globes)	17–21 points in the Optimize Energy Performance credit category (LEED) or 80–100 points in the Energy Performance section (Green Globes)
Silver level or 2 globes	25 percent abatement for 5 years	25 percent abatement for 5 years	25 percent abatement for 5 years	25 percent abatement for 5 years	25 percent abatement for 5 years	25 percent abatement for 5 years
Gold level or 3 globes	25 percent abatement for 5 years	30 percent abatement for 5 years	30 percent abatement for 5 years	30 percent abatement for 5 years	30 percent abatement for 5 years	30 percent abatement for 5 years
Platinum level or 4 globes	25 percent abatement for 5 years	30 percent abatement for 5 years	35 percent abatement for 5 years	35 percent abatement for 5 years	35 percent abatement for 5 years	35 percent abatement for 5 years

property with a maximum of 20 EV charger rebates available for each business location or multi-residential property. For example:

- 3 parking spaces = 1 EV charge rebate
- 13 parking spaces = 2 EV charger rebates
- 23 parking spaces = 3 EV charge rebates
- 33 parking spaces = 4 EV charger rebates[31]

10.5   Energy: Non-fiscal Tools

Cities and states in the Colorado River Basin have implemented a variety of energy-related non-fiscal tools that create interdependencies and synergies between the nexus systems while reducing trade-offs between the systems in the development of a green economy.

Power Content Label

The State of California's Energy Commission has created the Power Content Label, which provides consumers with reliable information about the energy resources used to generate electricity, enabling the ability to compare the power content of one electric service provider with that of others. The label provides information about the energy resources used to generate electricity that is put into the power grid as well as the sources of that energy. All electricity suppliers are required to include the label in their advertisements sent to customers in the mail or over the Internet. Furthermore, electricity suppliers must send an annual update for the electricity service product that is purchased by customers by October 1 each year. The label contains three columns:

- *Column A (Energy resources)*: This lists the different energy resources that can be used to generate electricity including renewable and fossil fuel resources.
- *Column B (Power mix)*: This column displays the actual mix of electricity purchased by the customers' utility in a given year, broken out by resource type.

- *Column C (California power mix)*: This column displays the mix of resources used in California for a given year. This information provides a reference point for customers to compare their electricity retail supplier's resource mix to the overall resource mix of the state.[32]

ResourceSmart Advice for Non-profits

Denver's Department of Strategic Partnership's ResourceSmart for Nonprofits helps eligible non-profit organisations reduce their utility costs without having to come up with the initial investment. Eligible organisations, including food banks, shelters, and halfway houses, affordable housing, and social service organisations, receive advice on how to take advantage of a unique combination of grants, incentives, rebates, and financing options once they have made a commitment to reduce their energy and water usage.[33]

Energy Savings for Schools Program

The Colorado Energy Office's (CEO) Energy savings for Schools Program provides energy management and project implementation support to Colorado K-12 public schools with a focus on rural and low-income communities. The programme involves CEO working with over 24 schools per year to achieve measurable savings and create sustainable energy programmes. The programme involves schools receiving a free energy and water audit, technical assistance, and coaching and implementation support. In particular, schools receive:

- On-site energy and water audits from a team or expert
- Evaluation of renewable energy opportunities
- Technical support and energy coaching
- Implementation support and help identifying existing funding and financing options to get projects completed
- Recognition for a school's efforts and opportunities to engage students[34]

10.6   Food: Fiscal Tools

Cities and states in the Colorado River Basin have implemented a variety of food-related fiscal tools that create interdependencies and synergies between the nexus systems while reducing trade-offs between the systems in the development of a green economy.

Urban Agriculture Property Tax Incentive

In 2014, California passed the Urban Agriculture Incentive Zones Act to increase the use of privately owned, vacant land for urban agriculture and improve land security for urban agricultural projects. The legislation enables city governments to designate areas within their boundaries as 'Urban Agriculture Incentive Zones'. In these zones landowners who commit their land to agricultural use for at least five years will receive a reduction in their property taxes. In particular, the property's tax assessment will be based on the agricultural value of the land rather than the market-rate value of the land. The law nonetheless comes with a range of restrictions and guidelines in its implementation including:

- Urban agricultural incentive zones can be established only in areas that fall within a designated urban area of 250,000 people or more.
- Projects must be on land that is at least 0.1 acre in size and no larger than three acres.
- Land must be completely dedicated towards commercial or non-commercial agricultural use.
- Land must not have any dwellings and only have physical structures that support agricultural production onsite.
- If a landowner breaks the five-year contract they are obliged to repay the tax benefits that they received.[35]

Specialty Crop Block Grant Program: Environmental Stewardship and Conservation

The California Department of Food and Agriculture (CDFA) awards competitive Specialty Crop Block Grant Program funds between

$50,000 and $450,000 to environmental stewardship and conservation projects that solely enhance the competitiveness of California specialty crops including fruits, vegetables, tree nuts, dried fruits, and horticulture and nursery crops. As part of the programme awards are given for projects that focus on enhancing soil health and conservation of agricultural land, water, habitat, and biodiversity as well as projects that enhance the specialty crops' contribution to climate change adaptation and/or mitigation. To be eligible for the grant in this section, projects must address at least one of the following programme priorities:

- Develop strategies and tools to enable specialty crop growers to adapt to climate change by reducing GHG emissions and sequestering carbon
- Develop innovations in water use efficiency and drought resilience
- Improve soil health and nutrient management
- Expand organic and sustainable production practices[36]

Sustainable Agricultural Lands Conservation Program

California's Sustainable Agricultural Lands Conservation Program provides funding for projects that protect at-risk agricultural lands from conversion to more GHG-intensive land uses, for example, urban or rural residential development, in order to promote growth within existing jurisdictions, ensure open space remains available, and support a healthy agricultural economy and resulting food security. There are two categories of grants available:

- Strategy and outcome grants support cities and counties with the development of local and regional land use policies and strategies that protect critical agricultural land, with $2.5 million available for strategy grants, with each individual grant limited up to $250,000 each (summarised in Table 10.5).[37]
- Agricultural conservation easements grants are provided to permanently protect the croplands and rangelands of willing landowners that are at risk of conversion, with $37.5 million available for property tax easement grants. The grants do not have a maximum dollar figure;

Table 10.5 Sustainable Agricultural Lands Conservation Program strategy and outcome grant

Strategy	Quantifiable outcomes
Establish an agricultural land mitigation programme	Agricultural conservation easement(s)
Establish an agricultural conservation easement purchasing programme	Agricultural conservation easement(s)
Adoption of urban limit line or urban growth boundary	Zoning ordinances that effectively eliminate growth in a project geographic area
Increase zoning minimum for designated strategic agricultural areas	Zoning ordinances that effectively eliminate growth in the project geographic area
Adoption of an agricultural greenbelt and implementation agreement	Results in both agricultural conservation easement(s) and zoning ordinances that effectively eliminate growth in the project geographic area

instead proposals will be rated relative to one another to maximise conservation outcomes with available funding.[38,39]

CalRecycle's Organics Grant Program

The California Department of Resources Recycling and Recovery (CalRecycle) offers an Organics Grant Program to lower the overall GHG emissions by expanding the existing capacity or establishing new facilities in California to reduce the amount of California-generated waste including food materials. Applications can be from local governments (cities, counties), private or for-profit entities, solid waste service providers, state agencies, universities, and non-profit organisations. Projects eligible for funding—up to $3 million per project—include the construction, renovation, or expansion of facilities in California that compost, anaerobically digest, or use other similar processes to turn food materials into value-added products including bioenergy and biofuel. Projects may also apply for a food waste prevention component to be included in the project that rescues edible food from becoming waste that is normally destined for landfills. All projects selected for funding must be able to demonstrate permanent, annual, and measurable reductions in GHG emissions from California-generated food materials and increases in quantity of California-generated food materials that are diverted from landfills.[40]

Utah's Low-Cost Agriculture Loan Programmes

Utah's Department of Agriculture and Food is responsible for loans that help the agricultural community and others achieve various worthwhile goals for productivity, efficiency, and environmental benefits for the people of Utah.

Utah's Agriculture Resource Development Loan Programme

Utah's Department of Agriculture and Food's Agriculture Resource Development Loan programme (ARDL) provides low-interest loans to farmers and ranchers to conserve soil and water, increase agricultural yield, maintain and improve water quality, conserve and improve wildlife habitat, prevent flooding, develop on-farm energy projects, and mitigate damages resulting from natural disasters (e.g. flooding, drought). The loans, made available through the Utah Conservation Commission in cooperation with the Department, are made for a maximum of 15 years at 2.5–3 percent with a one-time administrative fee of 4 percent. The loan specifics are in Table 10.6.[41]

Rural Rehabilitation Loan Program

The purpose of the Rural Rehabilitation Loan Program is to help financially troubled producers to stay in business, to assist beginning farmers in obtaining

Table 10.6   Utah's Agriculture Resource Development Loan programme

Loan amount	Interest rate	Loan terms	Max loan to value
Less than $52,000	3%	7 years or 15 years	70%
$52,001–$103,999	2.75%	15 years	70%
$104,000 or more	2.5%	15 years	70%

farm property, and to provide financing for transfer of agriculture properties from one generation to another. This loan is essentially a loan-of-last
resort requiring all applicants be declined by conventional commercial
lenders. The term of the loan is up to a maximum of ten years with interest rates 5 percent or less.[42]

Colorado's Grants for Energy Efficiency and Renewable Energy Projects

The Colorado Department of Agriculture, in partnership with the CEO,
is seeking applicants for agricultural energy efficiency and renewable
energy projects. The total amount of assistance for FY 2017 is $250,000
with funding available to Colorado agricultural irrigators, dairies, greenhouses, nurseries, and cold storage facilities with each project able to
receive up to $25,000 each. To be eligible, applicants must be enrolled in
the Colorado Agricultural Energy Efficiency Program and complete
either an energy audit to receive funding for energy efficiency projects or
complete a preliminary site assessment and technical support report to
receive funding for renewable energy projects.[43]

10.7   Food: Non-fiscal Tools

Cities and states in the Colorado River Basin have implemented a variety
of food-related non-fiscal tools that create interdependencies and synergies between the nexus systems while reducing trade-offs between the
systems in the development of a green economy.

Colorado Food Systems Advisory Council

The Colorado Food Systems Advisory Council (COFSAC) is a legislatively mandated, volunteer-based, 15-member body of state agencies and
food system stakeholders that advances recommendations to strengthen
healthy food access for all Coloradans through Colorado agriculture and
local food systems and economies. The Council does this by identifying

and using existing studies relevant to the food system, and developing recommendations that promote the building of a robust, resilient, and long-term local food economy. Currently the COFSAC is expanding access to locally grown food through farmers' markets and direct marketing of Colorado agricultural products.[44]

Utah Agriculture Certificate of Environmental Stewardship

Utah's Agriculture Certificate of Environmental Stewardship (ACES) programme helps farmers and ranchers of all sizes evaluate their entire operation to make management decisions that sustain agricultural viability, protect natural resources, support environmentally responsible agricultural production practices, and build a positive public opinion. To be eligible for certification, producers must complete three comprehensive steps:

1. Document completion of education modules relating to pesticide training, nutrient management, water management, and fertiliser application among others.
2. Complete a detailed application to evaluate on-farm risk.
3. Participate in an on-farm inspection to verify programme requirements applicable to state and federal environmental regulations. To retain the ACES, a participant must repeat all three steps including an inspection every five years.[45]

Colorado Dairy and Irrigation Efficiency Program

The Colorado Dairy and Irrigation Efficiency Program was launched in 2015 by the CEO to provide—via third-party contractors—free energy audits and technical support services to a minimum of 80 producers annually. CEO's contractor will serve as a point-of-contact for the programme and assist producers in selecting and implementing cost-effective improvements that reduce energy use, environmental impacts, and producer operator costs.[46]

10.8   Case Study Summary

Cities and states in the Colorado River Basin have implemented a variety of initiatives to create interdependencies and synergies between the nexus systems while reducing trade-offs between the systems in the development of a green economy.

General

- San Diego's Green Building Incentive Program offers fast turn-around times and reduced permit fees for projects that use resource-efficient construction materials and incorporate water and energy efficiency in the designs.

Water

- San Francisco offers grey water subsidies and rebates to encourage the installation of systems that divert laundry water to the garden. For large-scale grey water installations that require a permit, customers can receive a rebate to help cover the permit costs.
- Los Angeles offers large non-domestic water customers financial assistance for the installation of equipment and products that demonstrate water savings, with the incentive amount received based on the water savings accomplished by the project.
- Tucson requires all large non-domestic customers to conduct water audits and develop water conservation plans when a Stage 2 drought is declared. To help customers prepare for a drought, Tucson offers free water audits to help customers meet these requirements as well as rebates for installing water-efficient devices.
- California offers a water-energy grant to local agencies as well as non-profits to implement large-scale projects that enhance water and energy efficiency of both domestic and non-domestic water users.

- San Francisco's pilot community garden irrigation metering grant enables urban farms, including demonstration gardens, to track their water usage.
- South Nevada provides financial incentives for businesses to implement water efficiency capital expenditure projects that reduce water usage as well as wastewater flows.
- California is providing assistance to agricultural producers to implement irrigation systems that also reduce GHG emissions through energy efficiencies.

Energy

- Los Angeles' Custom Performance Program offers rebates for buildings or facilities to install various energy efficiency measures.
- Denver provides energy efficiency upgrades to non-profit organisations serving low-income clients. This reduces their operational costs ensuring more of their funding can be allocated towards programming.
- Nevada offers a green tax abatement programme in which energy-efficient building owners can receive property tax relief if their buildings meet various green building standards.
- Los Angeles offers a rebate for commercial customers to install EV charging stations at workplaces around the city.
- The State of California requires that all electricity suppliers disclose to customers the energy sources used to generate their electricity. To ensure ease of information the state's Power Content Label provides this information in an easy-to-read format enabling consumers to compare the power content of multiple electricity suppliers.
- Denver provides non-profits with advice on how to reduce energy costs and information on grants and other forms of financing available to reduce energy, as well as water, usage.

- Colorado is working with a selection of schools mainly in rural and low-income communities to implement energy management projects. As part of the programme schools receive free energy and water audits as well as technical advice.

Food

- California's Urban Agriculture Incentive Zone Act enables cities to offer property tax rebates for landowners who commit their land to urban agriculture.
- California is awarding grants to environmental projects that enhance the competitiveness of speciality crops with projects eligible for funding needing to address issues including water conservation and climate mitigation.
- California's sustainable land management programme provides funding to protect at-risk agricultural land from conversions to more GHG-intensive uses, for instance, urban or rural residential development.
- To reduce the amount of food waste entering landfill, California offers an organics grant to help entities develop or expand facilities that can turn food waste into bioenergy and biofuel.
- Utah is offering low-interest loans to farmers and ranchers for projects including improving agricultural yield, improving water quality, and developing on-farm energy projects.
- Colorado provides grants for on-farm energy efficiency and renewable energy projects.
- To enhance access to healthy food, Colorado's legislatively mandated multi-stakeholder advisory council identifies barriers to the current food system and develops recommendations to promote the building of a resilient, local food economy, including the development of

farmers' markets and direct marketing of locally produced agricultural products.

- Utah offers an environmental stewardship certificate for agricultural producers aiming to enhance their agricultural output while protecting natural resources. Part of the certificate involves producers completing education modules that include nutrient management and fertiliser application.
- Colorado provides free energy audits and technical advice to dairy producers that includes selecting and implementing cost-effective improvements that reduce energy and environmental impacts as well as lower operating costs.

Overall, cities and states in the Colorado River Basin use a variety of fiscal and non-fiscal tools to create interdependencies and synergies between the nexus systems while reducing trade-offs between the systems in the development of a green economy. These tools are summarised in Table 10.7.

Table 10.7 Colorado River Basin case summary

Tool	Tool type	Policy title	Description	WEF sectors addressed
Fiscal	Market-based instruments and pricing	Agriculture Resource Development Loan	Utah provides low-interest loans to farmers and ranchers to protect agricultural land and water and develop on-farm energy projects.	Food, energy, water
	Environmental taxes	Green building tax abatement	Nevada offers a tax abatement for new buildings or renovated buildings meeting LEED or Green Globes standard.	Water, energy
		Urban agriculture property tax incentive	California's Urban Agriculture Incentive Zone Act enables cities to offer property tax rebates for landowners who commit their land to urban agriculture.	Food
	Financial incentives	Green Building Incentive Program	San Diego County offers an expedited process, along with a fee reduction, for new and renovated green buildings.	Energy, water
		Residential grey water incentives	San Francisco offers subsidies and rebates for the installation of grey water systems.	Water
		Water Technical Assistance Program	LA's programme offers large water users funding for the installation of water-efficient equipment and products that demonstrate water savings.	Water
		Tucson's Audit Program	Tucson offers rebates for large water users to act on water-saving recommendations as part of their water audit.	Water

(continued)

Table 10.7 (continued)

Tool	Tool type	Policy title	Description	WEF sectors addressed
		Water-Energy Grant Program	California's programme offers funding for local agencies, joint power authorities, and non-profits initiating water-energy-saving programmes/projects.	Water, energy
		Community garden irrigation meter grant	San Francisco's grant is for the installation of a dedicated urban agriculture irrigation scheme and meter to track irrigation water use.	Water, food
		Water Efficient Technologies programme	Southern Nevada's programme provides incentives for businesses to efficiently use water and reduce wastewater entering the sanitary system.	Water
		State Water Efficiency Enhancement Program	California's programme provides grants for the installation of water-efficient irrigation on agricultural land.	Water, food, energy
		Custom Performance Program	LA's programme offers rebates for the installation of energy efficiency measures in buildings or facilities.	Energy
		Nonprofit Energy Efficiency Program	Denver offers funding for energy audits and upgrades to non-profits serving low-income clients.	Energy
		Electric vehicle charger commercial rebate	LA offers a commercial EV charger rebate to expand the city's EV charging infrastructure.	Energy

(continued)

Table 10.7 (continued)

Tool	Tool type	Policy title	Description	WEF sectors addressed
		Speciality crop block grant	California offers funding to enhance soil health, water conservation, and biodiversity of land used for specialty crops.	Food, water
		Organics grant programme	California is providing grant funding for projects that turn food waste into value-added products including bioenergy and biofuel.	Food, energy
		On-farm energy efficiency and renewable energy	Colorado's programme is providing funding for agriculture-related renewable energy and energy efficiency projects.	Energy, water
	Payment for ecosystem services	Sustainable Agricultural Lands Conservation Programme	California provides funding for projects that protect at-risk agricultural lands from conversion to urban development with funding available to permanently protect the croplands.	Food

(continued)

Table 10.7 (continued)

Tool	Tool type	Policy title	Description	WEF sectors addressed
Non-fiscal	Regulations	Tucson's Audit Program	During droughts, all large non-domestic water users need to conduct a self-audit of water use and develop a conservation plan.	Water
	Standards and mandatory labelling	Power content label	California mandates that all retail electricity suppliers must disclose to consumers the energy resources used.	Energy
	Public education and skills development	Energy Savings for Schools	Colorado provides support to schools for implementing sustainable energy programmes. This involves technical assistance and coaching.	Energy, water
	Stakeholder participation	Food Systems Advisory Council	Colorado's legislatively mandated volunteer council makes recommendations on strengthening healthy food access.	Food
	Information and awareness-raising	ResourceSmart advice	Denver provides advice for non-profits seeking to reduce their water and energy use.	Water, energy
		Agriculture Certificate of Environmental Stewardship	The programme helps farmers and ranchers of all sizes evaluate their entire operation to make management decisions that sustain agricultural viability.	Food, water
		Colorado Dairy and Irrigation Efficiency Program	The programme provides—via third-party contractors—free energy audits and technical support services.	Food, energy

Notes

1. Groves, D., Jordan R. Fischbach, Evan Bloom, Debra Knopman and Ryan Keefe. 2013. Adapting to a changing Colorado River: Making future water deliveries more reliable through robust management strategies. Available: http://www.rand.org/pubs/research_reports/RR242.html.
2. USGS. 2017. *Colorado River Basin focus area study* [Online]. Available: https://water.usgs.gov/watercensus/colorado.html.
3. Groves, D., Jordan R. Fischbach, Evan Bloom, Debra Knopman and Ryan Keefe. 2013. Adapting to a changing Colorado River: Making future water deliveries more reliable through robust management strategies. Available: http://www.rand.org/pubs/research_reports/RR242.html.
4. Ibid.
5. USBR. 2011. *Colorado River Basin water supply and demand study seeks input to help resolve projected future supply and demand imbalances* [Online]. Available: https://www.usbr.gov/newsroom/newsrelease/detail.cfm?RecordID=38644.
6. U.S. EPA. 2017. *Climate impacts in the Southwest* [Online]. Available: https://www.epa.gov/climate-impacts/climate-impacts-southwest.
7. Jon Unruh and Diana Liverman. 2016. *Changing water use and demand in the Southwest* [Online]. Available: https://geochange.er.usgs.gov/sw/impacts/society/water_demand/.
8. Center for Water Policy. 2013a. Climate change impacts on agriculture in the Colorado River Basin. Available: http://uwm.edu/centerforwater-policy/wp-content/uploads/sites/170/2013/10/Colorado_Agriculture_Final.pdf.
9. Los Alamos National Laboratory. 2016. *Energy-water challenge emerges in Colorado River flows* [Online]. Available: http://www.lanl.gov/discover/news-release-archive/2016/March/03.22-colorado-river-flows.php.
10. Tidwell, V. C., Dale, L., Franco, G., Averyt, K., Wei, M., Kammen, D. M., Nelson, J. H. & Barnhart, A. 2013. Energy: Supply, demand, and impacts. *Assessment of Climate Change in the Southwest United States: A report prepared for the National Climate Assessment.* Washington D.C.: Island Press.
11. Center for Water Policy. 2013b. Climate change impacts on hydropower in the Colorado River Basin. Available: http://uwm.edu/centerforwater-policy/wp-content/uploads/sites/170/2013/10/Colorado_Energy_Final.pdf.

12. Tidwell, V. C., Dale, L., Franco, G., Averyt, K., Wei, M., Kammen, D. M., Nelson, J. H. & Barnhart, A. 2013. Energy: Supply, demand, and impacts. *Assessment of Climate Change in the Southwest United States: A report prepared for the National Climate Assessment.* Washington D.C.: Island Press.
13. Pacific Institute. 2013. Water to supply the land: Irrigated agriculture in the Colorado River Basin. Available: http://pacinst.org/app/uploads/2013/05/pacinst-crb-ag.pdf.
14. Center for Water Policy. 2013a. Climate change impacts on agriculture in the Colorado River Basin. Available: http://uwm.edu/centerforwaterpolicy/wp-content/uploads/sites/170/2013/10/Colorado_Agriculture_Final.pdf.
15. San Diego County. 2017. *The Green building program* [Online]. Available: http://www.sandiegocounty.gov/pds/greenbuildings.html.
16. Ibid.
17. San Francisco Public Utilities Commission. 2017a. *Graywater* [Online]. Available: http://sfwater.org/index.aspx?page=100.
18. LADWP. 2013. Water conservation technical assistance program. Available: https://www.ladwp.com/ladwp/faces/ladwp/aboutus/a-water/a-w-conservation/a-w-c-customwaterprojectstap?_afrLoop=756416003347219&_afrWindowMode=0&_afrWindowId=null#%40%3F_afrWindowId%3Dnull%26_afrLoop%3D756416003347219%26_afrWindowMode%3D0%26_adf.ctrl-state%3Drso2jrezy_54.
19. City of Tucson. 2017. *Tucson audit program* [Online]. Available: https://www.tucsonaz.gov/water/tucson-audit-program.
20. City of Tucson. 2016. Tucson audit program policy and rebate application instructions. Available: https://www.tucsonaz.gov/files/water/docs/TAP_policy_instructions-9.1.2016.pdf.
21. California Department of Water Resources. 2016b. *Water-energy grant program* [Online]. Available: http://www.water.ca.gov/waterenergygrant/guidlinepsp.cfm.
22. California Department of Water Resources. 2016a. 2016 Water-energy grant program: Guidelines and proposal solicitation package. Available: http://www.water.ca.gov/waterenergygrant/docs/Final%202016_WE_GL_PSP_September2016.pdf.
23. San Francisco Public Utilities Commission. 2017b. *Urban agriculture and community grants* [Online]. Available: http://sfwater.org/index.aspx?page=469.

24. Southern Nevada Water Authority. 2017. *WET Custom technologies* [Online]. Available: https://www.snwa.com/biz/rebates_wet_custom. html.
25. California Department of Food and Agriculture. 2016. 2016 State water efficiency and enhancement program. Available: https://www.cdfa. ca.gov/oefi/sweep/docs/2016SWEEP-RndII-RequestGrantApp.pdf.
26. LADWP. 2017b. *Custom performance program* [Online]. Available: https://www.ladwp.com/ladwp/faces/ladwp/commercial/c-savemoney/c-sm-rebatesandprograms/c-sm-rp-cpp?_afr-Loop=757738576161541&_afrWindowMode=0&_afrWindowId=null#%40%3F_afrWindowId%3Dnull%26_afrLoop%3D757738576161541%26_afrWindowMode%3D0%26_adf.ctrl-state%3Drso2jrezy_80.
27. City and County of Denver. 2017. *Nonprofit energy efficiency support* [Online]. Available: https://www.denvergov.org/content/denvergov/en/human-rights-and-community-partnerships/our-offices/strategic-partnerships/energy-efficiency.html.
28. Nevada Governor's Office of Energy. 2017. *Green building tax abatements* [Online]. Available: http://energy.nv.gov/Programs/Green_Building_Tax_Abatements/.
29. Nevada Governor's Office of Energy. 2016. NAC 701A.280. Available: http://energy.nv.gov/uploadedFiles/energynvgov/content/Programs/NAC701A.280.pdf.
30. Ibid.
31. LADWP. 2017a. *Charge up L.A.!* [Online]. Available: https://www.ladwp.com/ladwp/faces/ladwp/residential/r-gogreen/r-gg-driveelectric;jsessionid=1DPMYzVQkLJJqvLN6Dxnx7SGhtsbQpSq6pxlm4vJnc54rV2mJ3Yw!-801040537?_afrLoop=158291810020083&_afrWindowMode=0&_afrWindowId=null#%40%3F_afrWindowId%3Dnull%26_afrLoop%3D158291810020083%26_afrWindowMode%3D0%26_adf.ctrl-state%3D1cmbpvmu4f_4.
32. California Energy Commission. 2017. *About the power content label* [Online]. Available: http://www.energy.ca.gov/pcl/power_content_label.html.
33. City and County of Denver. 2017. *Nonprofit energy efficiency support* [Online]. Available: https://www.denvergov.org/content/denvergov/en/human-rights-and-community-partnerships/our-offices/strategic-partnerships/energy-efficiency.html.

34. Colorado Energy Office. 2017. *Energy savings for schools* [Online]. Available:https://www.colorado.gov/pacific/energyoffice/energy-savings-schools.
35. California State Board of Equalization. 2013. *Urban Agriculture Incentive Zones Act* [Online]. Available: https://www.boe.ca.gov/proptaxes/uaincentivezone.htm.
36. California Department of Food and Agriculture. 2017. 2017 specialty crop block grant program. Available: https://www.cdfa.ca.gov/Specialty_Crop_Competitiveness_Grants/pdfs/2017SCBGP_RequestForConcept Proposals.pdf.
37. California Strategic Growth Council. 2016. Sustainable agricultural lands conservation. Available: http://sgc.ca.gov/pdf/AHSC_SALC%20 Fact%20Sheet_FINAL.pdf.
38. California Strategic Growth Council. 2017. *Sustainable agricultural lands conservation (SALC) program overview* [Online]. Available: http://sgc.ca.gov/Grant-Programs/SALC-Program.html.
39. California Strategic Growth Council. 2015. California sustainable agricultural lands conservation program final program guidelines. Available: http://www.conservation.ca.gov/dlrp/Documents/FY2015%20 SALCP%20Final%20Guidelines_12.18.2015.pdf.
40. CalRecycle. 2015. *Notice of funds available: Organics grant program (FY 2014–2015)* [Online]. Available: http://www.calrecycle.ca.gov/Climate/ GrantsLoans/Organics/FY201415/default.htm.
41. Utah Department of Agriculture and Food. 2017a. *Agriculture loans* [Online]. Available: http://ag.utah.gov/markets-finance/agriculture-loans.html.
42. Utah Department of Agriculture and Food. Low-cost loans. Available: http://ag.utah.gov/markets-finance/agriculture-loans/29-conservation-and-environmental/loans/204-low-cost-loans.html.
43. Colorado Department of Agriculture. 2017. *2017 RCCP Agricultural energy program grant funding* [Online]. Available: https://www.colorado. gov/pacific/agconservation/2017-energy-project-funding.
44. Colorado Food Systems Advisory Council. 2017. *About COFSAC* [Online]. Available: http://www.cofoodsystemscouncil.org/.
45. Utah Department of Agriculture and Food. 2017b. *Why are we here?* [Online]. Available: http://ag.utah.gov/aces/index.html.
46. Colorado Energy Office. 2016. Colorado dairy and irrigation efficiency participant application. Available: https://www.colorado.gov/pacific/ energyoffice/atom/44691.

References

California Department of Food and Agriculture. 2016. 2016 State water efficiency and enhancement program. https://www.cdfa.ca.gov/oefi/sweep/docs/2016SWEEP-RndII-RequestGrantApp.pdf
———. 2017. 2017 specialty crop block grant program. https://www.cdfa.ca.gov/Specialty_Crop_Competitiveness_Grants/pdfs/2017SCBGP_RequestForConceptProposals.pdf
California Department of Water Resources. 2016a. 2016 water-energy grant program: Guidelines and proposal solicitation package. http://www.water.ca.gov/waterenergygrant/docs/Final%202016_WE_GL_PSP_September2016.pdf
———. 2016b. *Water-energy grant program* [Online]. http://www.water.ca.gov/waterenergygrant/guidlinepsp.cfm
California Energy Commission. 2017. *About the power content label* [Online]. http://www.energy.ca.gov/pcl/power_content_label.html
California State Board of Equalization. 2013. *Urban Agriculture Incentive Zones Act* [Online]. https://www.boe.ca.gov/proptaxes/uaincentivezone.htm
California Strategic Growth Council. 2015. California sustainable agricultural lands conservation program final program guidelines. http://www.conservation.ca.gov/dlrp/Documents/FY2015%20SALCP%20Final%20Guidelines_12.18.2015.pdf
———. 2016. Sustainable agricultural lands conservation. http://sgc.ca.gov/pdf/AHSC_SALC%20Fact%20Sheet_FINAL.pdf
———. 2017. *Sustainable agricultural lands conservation (SALC) program overview* [Online]. http://sgc.ca.gov/Grant-Programs/SALC-Program.html
CalRecycle. 2015. *Notice of funds available: Organics grant program (FY 2014–2015)* [Online]. http://www.calrecycle.ca.gov/Climate/GrantsLoans/Organics/FY201415/default.htm
Center for Water Policy. 2013a. Climate change impacts on agriculture in the Colorado River Basin. http://uwm.edu/centerforwaterpolicy/wp-content/uploads/sites/170/2013/10/Colorado_Agriculture_Final.pdf
———. 2013b. Climate change impacts on hydropower in the Colorado River Basin. http://uwm.edu/centerforwaterpolicy/wp-content/uploads/sites/170/2013/10/Colorado_Energy_Final.pdf
City and County of Denver. 2017. *Nonprofit energy efficiency support* [Online]. https://www.denvergov.org/content/denvergov/en/human-rights-and-community-partnerships/our-offices/strategic-partnerships/energy-efficiency.html

City of Tucson. 2016. Tucson audit program policy and rebate application instructions. https://www.tucsonaz.gov/files/water/docs/TAP_policy_instructions-9.1.2016.pdf

———. 2017. *Tucson audit program* [Online]. https://www.tucsonaz.gov/water/tucson-audit-program

Colorado Department of Agriculture. 2017. *2017 RCCP Agricultural energy program grant funding* [Online]. https://www.colorado.gov/pacific/agconservation/2017-energy-project-funding

Colorado Energy Office. 2016. Colorado dairy and irrigation efficiency participant application. https://www.colorado.gov/pacific/energyoffice/atom/44691

———. 2017. *Energy savings for schools* [Online]. https://www.colorado.gov/pacific/energyoffice/energy-savings-schools

Colorado Food Systems Advisory Council. 2017. *About COFSAC* [Online]. http://www.cofoodsystemscouncil.org/

Groves, D., Jordan R. Fischbach, Evan Bloom, Debra Knopman, and Ryan Keefe. 2013. Adapting to a changing Colorado River: Making future water deliveries more reliable through robust management strategies. http://www.rand.org/pubs/research_reports/RR242.html

LADWP. 2013. Water conservation technical assistance program. https://www.ladwp.com/ladwp/faces/ladwp/aboutus/a-water/a-w-conservation/a-w-c-customwaterprojectstap?_afrLoop=756416003347219&_afrWindowMode=0&_afrWindowId=null#%40%3F_afrWindowId%3Dnull%26_afrLoop%3D756416003347219%26_afrWindowMode%3D0%26_adf.ctrl-state%3Drso2jrezy_54

———. 2017a. *Charge up L.A.!* [Online]. https://www.ladwp.com/ladwp/faces/ladwp/residential/r-gogreen/r-gg-driveelectric;jsessionid=1DPMYzVQkLJJqvLN6Dxnx7SGhtsbQpSq6pxlm4vJnc54rV2m-J3Yw!-801040537?_afrLoop=158291810020083&_afrWindowMode=0&_afrWindowId=null#%40%3F_afrWindowId%3Dnull%26_afrLoop%3D158291810020083%26_afrWindowMode%3D0%26_adf.ctrl-state%3D1cmbpvmu4f_4

———. 2017b. *Custom performance program* [Online]. https://www.ladwp.com/ladwp/faces/ladwp/commercial/c-savemoney/c-sm-rebatesandprograms/c-sm-rp-cpp?_afrLoop=757738576161541&_afrWindowMode=0&_afrWindowId=null#%40%3F_afrWindowId%3Dnull%26_afrLoop%3D757738576161541%26_afrWindowMode%3D0%26_adf.ctrl-state%3Drso2jrezy_80

Los Alamos National Laboratory. 2016. *Energy-water challenge emerges in Colorado River flows* [Online]. http://www.lanl.gov/discover/news-release-archive/2016/March/03.22-colorado-river-flows.php

Nevada Governor's Office of Energy. 2016. NAC 701A.280. http://energy.nv.gov/uploadedFiles/energynvgov/content/Programs/NAC701A.280.pdf

———. 2017. *Green building tax abatements* [Online]. http://energy.nv.gov/Programs/Green_Building_Tax_Abatements/

Pacific Institute. 2013. Water to supply the land: Irrigated agriculture in the Colorado River Basin. http://pacinst.org/app/uploads/2013/05/pacinst-crb-ag.pdf

San Diego County. 2017. *The Green building program* [Online]. http://www.sandiegocounty.gov/pds/greenbuildings.html

San Francisco Public Utilities Commission. 2017a. *Graywater* [Online]. http://sfwater.org/index.aspx?page=100

———. 2017b. *Urban agriculture and community grants* [Online]. http://sfwater.org/index.aspx?page=469

Southern Nevada Water Authority. 2017. *WET Custom technologies* [Online]. https://www.snwa.com/biz/rebates_wet_custom.html

Tidwell, V.C., L. Dale, G. Franco, K. Averyt, M. Wei, D.M. Kammen, J.H. Nelson, and A. Barnhart. 2013. Energy: Supply, demand, and impacts. *Assessment of Climate Change in the Southwest United States: A Report Prepared for the National Climate Assessment.* Washington, DC: Island Press.

U.S. EPA. 2017. *Climate impacts in the Southwest* [Online]. https://www.epa.gov/climate-impacts/climate-impacts-southwest

Unruh, Jon, and Diana Liverman. 2016. *Changing water use and demand in the Southwest* [Online]. https://geochange.er.usgs.gov/sw/impacts/society/water_demand/

USBR. 2011. *Colorado River Basin water supply and demand study seeks input to help resolve projected future supply and demand imbalances* [Online]. https://www.usbr.gov/newsroom/newsrelease/detail.cfm?RecordID=38644

USGS. 2017. *Colorado River Basin focus area study* [Online]. https://water.usgs.gov/watercensus/colorado.html

Utah Department of Agriculture and Food. 2017a. *Agriculture loans* [Online]. http://ag.utah.gov/markets-finance/agriculture-loans.html

———. 2017b. *Why are we here?* [Online]. http://ag.utah.gov/aces/index.html

———. n.d. Low-cost loans. http://ag.utah.gov/markets-finance/agriculture-loans/29-conservation-and-environmental/loans/204-low-cost-loans.html

11

The Green Economy and the Water-Energy-Food Nexus in the Murray-Darling River Basin

Introduction

The Murray-Darling River Basin Plan provides a coordinated approach to water use across the Murray-Darling River Basin's four states (Queensland, New South Wales, Victoria, and South Australia) and the Australian Capital Territory (ACT). The Basin Plan, passed into law in November 2012, balances social, economic, and environmental demands on the Basin's water resources to ensure:

- Strong and vibrant communities with sufficient water of a suitable quality for drinking and domestic uses (including in times of drought), as well as for cultural and recreational purposes
- Productive and resilient industries that have a long-term confidence in their future, particularly for food and fibre production
- Healthy and diverse ecosystems with rivers regularly connected to their creeks, billabongs, and floodplains
- An increase in the amount of water for the environment of the Murray-Darling River Basin and ensuring sufficient water for all users

© The Author(s) 2018
R.C. Brears, *The Green Economy and the Water-Energy-Food Nexus*,
DOI 10.1057/978-1-137-58365-9_11

11.1   Water-Energy-Food Nexus Pressures

The Murray-Darling River Basin is experiencing a variety of water, energy, and food nexus pressures that are detrimental to the development of a green economy as described below through a variety of examples.

Water

In New South Wales (NSW), demand for the state's water is high, with urban water use having an impact on river flow in areas of significant population density. Water supplied to NSW cities and large towns has had an annual growth rate of 1.35 percent over the past seven years.[1] Regarding the impacts of climate change, modelling conducted under the Sydney Water Balance Project has found that climate change will decrease the annual rainfall and runoff in the inland catchments. Climate change is also likely to result in an increase in evaporation throughout the catchments with a 22 percent increase in pan-evaporation in inland catchments and a 9 percent increase in coastal catchments by 2070.[2] Similarly, Victoria is expected to experience a reduction in average surface water by 2030. In Melbourne, the average long-term stream flow into water supply catchments could be reduced by up to 11 percent by 2020 and 35 percent by 2050.[3]

Energy

In NSW, final energy supplied to the end user, which includes secondary energy, for example, electricity, has been decreasing by nearly 3 percent per annum. However, the majority of fuel for electricity comes from non-renewable sources such as coal (82.3 percent). Meanwhile, renewable sources provide nearly 11 percent of the state's electricity production.[4] Similarly, while renewable energy sources including wind and solar power are playing an increasing role in Victoria's electricity generation, fossil fuels remain the major source of the state's electricity, currently accounting for more than 85 percent of the state's electricity generation.[5]

The heavy reliance on fossil fuels has environmental and health impacts including GHG emissions.[6]

Food

Agriculture is a major user of natural resources in NSW, with around 75 percent of land mass having an agricultural land use, and around 80 percent of all water extracted from regulated rivers is diverted to agricultural production.[7] Poor soil quality is impacting the production of food; for instance, in South Australia, about 40 percent of the area under broadacre agriculture (production of grains, oilseeds, and other crops including wheat, barley, and maize), have issues that limit soil productivity.[8] Climate change is likely to reduce agricultural productivity; for instance, in parts of Victoria average grain yields are likely to decrease by 10–20 percent by 2050.[9]

11.2 General Fiscal Tools to Reduce Water-Energy-Food Nexus Pressures

Cities, states and the territory in the Murray-Darling River Basin have implemented a variety of fiscal tools that create interdependencies and synergies between the nexus systems while reducing trade-offs between the systems in the development of a green economy.

ecoMarkets

ecoMarkets is a Victorian Government initiative that provides incentives for private landholders—who own 65 percent of Victoria's land—to manage their land in ways that conserve and enhance the environment. Landholders will be able to earn income from ecoMarkets if they are able to provide cost-effective environmental improvements. Willing buyers may be the government seeking to obtain environmental improvements on behalf of the Victorian public or private companies or individuals, for example, developers seeking to offset environmental damage through remediation elsewhere. The programme has three types of tenders: BushTender, EcoTender, and BushBroker.

BushTender

BushTender aims to improve the management of existing areas of native vegetation on private land. Landholders nominate their own bid price in a competitive tender and choose a range of actions to protect and enhance native vegetation, for example, fencing of native vegetation to exclude stock. Successful bids are ones that offer 'best value for money' in terms of native vegetation and biodiversity outcomes.

EcoTender

EcoTender expands the BushTender programme by including multiple environmental benefits. In addition to native vegetation, landholders' bids are evaluated on their potential improvements to river and estuary health. Under EcoTender, landholders put out for tender contracts that deliver several complementary benefits primarily through improved native vegetation management and re-vegetation on their properties. Successful bids contain activities that offer the best value for money to the community based on ecosystem outcomes, the significance of the environmental assets, and the cost.

BushBroker

BushBroker provides a system where native vegetation credits can be generated and traded, allowing landholders to provide credits on behalf of others. Landholders can provide native vegetation credits from their property by protecting and better managing remnant bushland through activities including fencing off stock. Credits can also be earned by re-vegetating previously cleared land with native plants and by protecting scattered paddock trees to encourage natural re-vegetation. Putting freehold land into conservation reserves can also earn credits.[10]

11.3 General Non-fiscal Tools to Reduce Water-Energy-Food Nexus Pressures

Cities, states and the territory in the Murray-Darling River Basin have implemented a variety of non-fiscal tools that create interdependencies and synergies between the nexus systems while reducing trade-offs between the systems in the development of a green economy.

Actsmart Schools

Actsmart Schools is a free, specialised programme that helps schools become sustainable by saving energy, conserving water, increasing recycling, protecting biodiversity, reducing greenhouse gases, and integrating sustainability into the curriculum. The programme provides support and/or training for schools to:

- Develop an environmental plan for the sustainable management of waste, water, energy, and school grounds
- Audit their water, waste, and energy
- Review their school grounds
- Review their curriculum
- Educate teachers, business managers, facilities managers, and students on sustainable practices

The programme also offers information and tools including:

- Curriculum units on energy, water, waste, biodiversity, and climate change
- An operational guide to becoming a sustainable school which provides best practices on managing energy, water, waste, and school ground biodiversity and how to integrate sustainability into the curriculum[11]

Sustainability Advantage Recognition Scheme

Sustainability Advantage is a business support service from the NSW Office of Environment and Heritage that aims to help organisations improve their environmental performance, reduce costs, and add value to their corporate reputation. The programme works with over 550 organisations in NSW and helps them identify and implement projects in practical areas including:

- Resource efficiency (energy, water, waste, and raw materials)
- Supply chain
- Staff engagement
- Carbon management

All types of organisations, ranging from medium and large businesses, not-for-profits, and government organisations, can participate in the programme. Members are grouped into 34 clusters (17 regional and 17 Sydney-based) around a geographic or business need. These clusters facilitate networking, peer-learning and support, and strategic projects. Participants come from a variety of industry sectors including manufacturing, commercial property, registered clubs, health care, transport, and education. Participants also benefit from attending masterclasses, workshops, and one-on-one support including green marketing, behaviour change for sustainability, measure to manage resources, and sustainable event management. To participate in the programme organisations must:

- Commit to participate for 18 months
- Complete the sustainability management diagnostic which helps management identify and prioritise environmental actions that minimise risk and create business opportunities
- Choose and undertake two or three of the seven Sustainability Advantage modules:
  - Vision, commitment, and planning
  - Resource efficiency
  - Staff engagement
  - Carbon management

- Supply chain management
- External stakeholder engagement
- Environmental risk and responsibility

- Network with other organisations of the same industry or region three or four times a year
- Report progress[12]

Recognition of Participation

The Sustainability Advantage recognition scheme rewards businesses for their commitment to sustainability and real improvements they make on their sustainability journey through participation in the programme. To be recognised as a Sustainability Advantage partner an organisation must demonstrate progress towards sustainable practices through active participation, leadership, commitment, and planning; internal and external engagement; and achievements. Upon joining the Sustainability Advantage programme organisations become a Member and have the ability to progress towards Bronze, Silver, Gold, or Platinum awards (Table 11.1).[13]

Table 11.1 Sustainability Advantage recognition award

Award	Description
Bronze	Recognises organisations that have been a member for at least 12 months and have demonstrated commitment to business sustainability.
Silver	Recognises organisations that have been Bronze for at least 12 months and can demonstrate significant environmental achievements.
Gold	Recognises organisations that have been Silver for at least two years and can demonstrate outstanding environmental achievement and leadership.
Platinum	Recognises organisations that have been Gold for at least three years and can demonstrate innovation, performance, and competitive advantage through practices that achieve 'net zero impact' on the environment. Platinum members also have to initiate a Platinum project, which is innovative and creates a transformational shift in an organisation's business model, product, service, or process.

ResourceSmart Schools

ResourceSmart Schools is a framework that helps schools embed sustainability into everything they do. By participating in the programme Victorian schools learn how to manage their energy, waste, water, and biodiversity. The programme enables participating schools to manage and track their resource use over time through ResourceSmart Schools online. They are also given access to tools and support so they can incorporate sustainability into their curriculum. To date more than 50 percent of Victorian schools and 400,000 students have:

- Diverted 15,000 cubic metres of waste from landfill, saving $6 million
- Reduced electricity consumption by 19,000 MWh and 34,000 tonnes of carbon dioxide, saving $6.7 million
- Reduced water consumption by 950 ML, saving $2.2 million
- Planted 5.5 million plants[14]

ResourceSmart School Awards

The ResourceSmart Schools Awards are held annually to recognise the most innovative and outstanding sustainability programmes and initiatives taking place in schools across Victoria. Through the ResourceSmart School Awards, students and teachers are rewarded for their creative actions that reduce energy, waste, and water; increase biodiversity; and how they engage the wider community on these issues. The awards are summarised in Table 11.2.[15]

Green Globe Awards

The Green Globe Awards recognise excellence in environmental sustainability in NSW. The awards are for businesses, government, community groups, and individuals who are leading the way in making NSW greener and more innovative. There are awards for organisations and individuals, Awards for Impact, and Best of the Best Awards, the details of which are summarised in Table 11.3.[16]

Table 11.2 ResourceSmart School Awards

Award	Description
Biodiversity School of the Year (Primary and Secondary)	Recognises schools that have initiated or influenced positive biodiversity changes within their school or broader community.
Community Leadership School of the Year (Primary and Secondary)	Recognises schools that have shown outstanding leadership by working with local groups in their community to improve environmental sustainability.
Energy School of the Year (Primary and Secondary)	Recognises schools that have initiated or influenced positive energy changes within their school or broader community.
Waste School of the Year (Primary and Secondary)	Recognises schools that have initiated or influenced positive waste changes within their school or broader community.
Water School of the Year (Primary and Secondary)	Recognises schools that have initiated or influenced positive water changes within their school or broader community.
Student Action Team of the Year (Primary and Secondary)	Recognises individuals and groups of students who have initiated or influenced positive sustainability changes in their school, local communities, and/or regional areas. Nominated teams stand out as leaders who have worked together to achieve outstanding results in sustainability.
Teacher of the Year (Primary and Secondary)	Recognises teachers who have initiated or influenced positive sustainability change within their school or broader community. Nominated teachers are standout teachers who have gone above and beyond their regular duties in their school's sustainability services.
Early Childhood Service of the Year	Recognises the early childhood service provider that implemented substantial sustainability initiatives.
School of the Year	Schools that enter three or more categories are eligible for the ResourceSmart School of the Year Award.

Awards for Organisations and Individuals

Nominations for awards for organisations and individuals that demonstrate excellence in all aspects of sustainability including environmental, social, and economic.

Table 11.3 New South Wales's Green Globe Awards

Award category	Award	Description
Awards for organisations and individuals	Business leadership award	For outstanding NSW businesses or corporations that have fully integrated environmental management and sustainable practices into their planning, strategy, and operations and/or service delivery to strengthen their commercial viability.
	Community leadership award	Recognises community leadership and commitment from non-profit organisations whose local sustainability initiatives have widespread benefits for NSW communities.
	Public sector leadership award	Recognises public sector organisations that have successfully integrated environmental management and sustainable practices into planning, operations, product, and/or service delivery.
	Sustainability champion award	Recognises individuals who have been instrumental in delivering environmental projects and demonstrating leadership in influencing and changing community or organisation views and/or driving change in sustainable practices. This is open to individuals 26 years and older living in NSW.
	Young sustainability champion award	Recognises and rewards a young individual for practical environmental solutions and promoting and engaging a community to improve its environmental issues. This award overall recognises a future environmental leader who is 25 years old and younger living in NSW.

(continued)

Table 11.3 (continued)

Award category	Award	Description
Awards for Impact	Built environment award	Recognises demonstrated excellence and innovation in designing, constructing, retrofitting, and managing of existing or new buildings, precincts, and tenancies. This may be for commercial, residential, heritage properties, and infrastructure projects in NSW.
	Natural environment award	Recognises leadership in protecting and enhancing natural ecosystems, including native flora and fauna, natural habitats, water resources, land systems, and soil and biodiversity.
	Resource efficiency award	Recognises excellence in integrated practical solutions for clean energy, water saving and conservation, waste avoidance, resource recovery, and/or recycling practices.
	Climate change leadership award	Recognises exceptional work and/or leadership in climate change mitigation and adaptation in NSW. This may be for areas including reducing GHG emissions, managing climate risks, empowering communities, businesses, and/or the environment to adapt to climate change, leading the emergence of the low-carbon economy and initiatives to influence policies, practices, and attitudes.
	Innovation award	Recognises outstanding new technology, design, or research in the areas of, for example, sustainable technology, new industry practices, products or services, and improving the ability to use, store, or save natural resources.
	Premier's award for sustainable excellence	Presented to the most outstanding nomination from across all award categories. The recipient is chosen for exemplary leadership, vision, innovation as well as determination in delivering outstanding benefits to the environment and economy.
Best of the Best Awards	Regional sustainability award	Recognises the outstanding achievements of an organisation, project, or individual who has overcome unique challenges in regional or rural areas to deliver innovative and successful projects.
	10-year sustainability achievement award	This award, by invitation only, honours Green Globe Awards prior to 2006 who can demonstrate at least ten years of consistent, long-term sustainable practices to achieve better environmental outcomes.

Awards for Impact

Awards for Impact are open to all entrants including businesses, individuals, NSW public sector, NSW local councils, community organisations, partnerships between business, non-government organisations, government and community groups, research groups, academics, and recipients of funding from the NSW Environmental Trust and other bodies.

Best of the Best Awards

The Best of the Best Award categories are not open for nominations; instead, award winners are selected by a judging panel from the finalists in each of the Green Globe award categories.

11.4   Water: Fiscal Tools

Cities, states and the territory in the Murray-Darling River Basin have implemented a variety of water-related fiscal tools that create interdependencies and synergies between the nexus systems while reducing trade-offs between the systems in the development of a green economy.

Water Rights Trading

Water markets allow water to flow where it can be used most productively. In the Murray-Darling River Basin, the water market is based on a 'cap and trade' system where the cap represents the total pool of water available for consumption. The available water is distributed to users via water rights that are administered by the basin states. There are two types of water rights traded:

1. Water access entitlements are rights to an ongoing share of the total amount of water available in a system.
2. Water allocations are the actual amount of water available under water access entitlements in a given season.

From which, there are two types of trades:

- A permanent trade is the trade of water entitlements, an entitlement trade.
- A temporary trade is the trade of water allocations, an allocation trade.[17]

During the year water is allocated against entitlements by state governments in response to factors including rainfall, inflows into storages, and how water in storage is managed by the Basin states. At the start of each water year (July 1), each Basin state makes water allocation announcements based on seasonal availability. In regulated systems, allocations are reviewed throughout the year. This provides users with certainty as to the water they will receive while allowing states to manage water availability through different climatic conditions. The Basin states are responsible for:

- Determining water allocations
- Developing policies and procedures
- Monitoring water use
- Developing water resource[18]

Meanwhile, the actual trading of water allocations and entitlements is conducted via water brokerage firms.

South Australia's Water Charges

South Australia has a statewide price for most water services. Irrespective of whether water users are in a metropolitan or regional area, all customers pay the same price per kilolitre (one kilolitre equals 1000 litres). Water charges are comprised of two parts:

1. A fixed charge for water supply
2. A variable charge for water use

Table 11.4 SA Water residential water charges

Tier	User charge per kilolitre	Price per litre	Indicative quarterly threshold (kilolitres)	Daily threshold (kilolitres)
1	$2.27	$0.00227	0–30	0–0.3288
2	$3.24	$0.00324	30–130	0.3288–1.4247
3	$3.51	$0.00351	Above 130	Above 1.4247

Residential Water Charges

Residential water uses pay a water supply charge of $71.60 per quarter and a tiered variable water charge summarised in Table 11.4,[19] while customers with recycled water for irrigation, gardens, or flushing toilets pay a set price of $2.04 per kilolitre, which is 90 percent of tier 1 water prices.

Commercial Water Charges

The water supply charge for commercial customers is based on their property value with a minimum charge of $71.60 per quarter. The quarterly rate in the dollar applied to commercial customers is $0.174750 per $1000 of property value. The water use charge for commercial customers is $3.24 per kilolitre.[20]

Community Sustainability Action Grants: Water

From 2016 onwards, the Queensland Government's Community sustainability action grant is distributing $12 million over the next three years to eligible individuals and community groups for innovative projects that seek to address climate change and conserve Queensland's natural and built environment. Grants of up to $25,000 will be available for land care, environmental community, and volunteer groups as well as non-profit groups proposing to undertake sustainable and environmental conservation projects. As part of the grant scheme awards will be made for initiatives that build sustainable communities. Projects funded will improve water and energy efficiency of community facilities to provide ongoing community benefit. Specific activities that will be funded include installation of water- and energy-efficient pumps and water infrastructure for irrigation, installation of renewable energy systems such as PV systems and small-scale wind power, and replacement of water-using devices and appliances with more efficient ones.[21]

11.5   Water: Non-fiscal Tools

Cities, states and the territory in the Murray-Darling River Basin have implemented a variety of water-related non-fiscal tools that create interdependencies and synergies between the nexus systems while reducing trade-offs between the systems in the development of a green economy.

Schools Water Efficiency Program

The Schools Water Efficiency Program assists Victorian schools in monitoring their water usage for leakages and educates students on water consumption. The programme provides schools with data loggers to continuously record water use. The data is uploaded to an interactive web portal that enables schools to use the data as part of their curriculum and help manage water use. The data loggers will identify high water usage and leaks, saving water and money. As part of the programme schools:

- Will have installed data loggers (maximum of two free)
- Are provided a web-based platform for delivering school-specific water consumption data in collaboration with classroom materials and the ability to report on school water consumption
- Can easily identify leaky pipes and high water use and identify faulty appliances on school premises as feedback will be provided to schools where leaks or unusual usage is detected[22]

Sydney Water Benchmarking Water Users

Sydney Water has developed simple benchmarks to help businesses compare their water use with other similar business types. To work out a business's benchmark, each business can follow a three-step process:

1. Determine their water use (kilolitres per year)
2. Identify their yearly key business activity indicator (KBAI), which is the measure of how much water an efficient business uses for each specific activity (examples of which are summarised in Table 11.5)[23]
3. Divide the kilolitres of water used each year by the KBAI

Table 11.5 Sydney Water benchmarks for various sectors

Sector	Rating	Benchmark
Aquatic leisure centres	Best practice	<20 litres per bather
	Efficient	20–40 litres per bather
	Fair	40–60 litres per bather
	Inefficient	>60 litres per bather
Commercial kitchens	Efficient	<35 litres per food cover
	Fair	35–45 litres per food cover
	Inefficient	>45 litres per food cover
Commercial office buildings and shopping centres	Best practice	0.77 kilolitres/m²/year (with cooling towers) 0.40 kilolitres/m²/year (without cooling towers)
	Efficient	0.84 kilolitres/m²/year (with cooling towers) 0.47 kilolitres/m²/year (without cooling towers)
	Fair	1.01 kilolitres/m²/year (with cooling towers) 0.64 kilolitres/m²/year (without cooling towers)
Hotels	Best practice	0.4 kilolitres per room per day
	Efficient	0.4–0.45 kilolitres per room per day

11.6   Energy: Fiscal Tools

Cities, states and the territory in the Murray-Darling River Basin have implemented a variety of energy-related fiscal tools that create interdependencies and synergies between the nexus systems while reducing trade-offs between the systems in the development of a green economy.

Sustainability Victoria Business—Boosting Productivity Grants

Sustainability Victoria offers energy assessment grants for SMEs and energy efficiency capability grants for industrial associations, business networks, or registered training organisations to develop energy efficiency knowledge and skills in SMEs or service providers with the overall aim of helping 1000 SMEs with energy efficiency skills and connect with them with expert advice to help reduce energy costs and GHG emissions. The specific details of the two grants are as follows:

- *The energy assessment grant*: Covers 50 percent of the cost of an energy assessment with grants offered at two funding levels: a basic energy assessment up to $2000 and a detailed energy assessment up to $6000, with access to a $3000 assessment recommendation implementation bonus.
- *The energy efficiency capability grant*: Competitive grants up to $150,000 are available for projects that deliver energy efficiency capabilities to either SMEs or service providers with projects expected to deliver lasting change.[24, 25]

Business Energy and Water Programme

This programme offers rebates up to $5000 for ACT businesses wishing to upgrade to more energy, and water, efficient technologies, including lighting, heating and cooling, refrigeration, toilets, and/or tapware. To qualify for the rebate businesses must be operating in the ACT, have electricity bills up to $20,000 per year and/or employ a maximum of ten full-time staff. If accepted an Actsmart energy assessor will conduct an overview of energy and water use in the business and provide a detailed action plan with recommended energy and water improvements. Once the business has implemented the recommended improvements, they can claim back 50 percent (up to $5000) of the costs of the improvements, after which the business will receive a certificate of participation and recognition as an Actsmart business, enabling them to promote their environmentally responsible stance to staff, customers, and suppliers.[26]

11.7 Energy: Non-fiscal Tools

Cities, states and the territory in the Murray-Darling River Basin have implemented a variety of energy-related non-fiscal tools that create interdependencies and synergies between the nexus systems while reducing trade-offs between the systems in the development of a green economy.

Sustainability Victoria's Smarter Choice Retail Programme

Sustainability Victoria's Smarter Choice retail programme provides practical energy and water efficiency information to ease increasing living costs by saving consumers money on energy and water bills. Smarter Choice also informs households on how to recycle or dispose of their old appliances. Over 400 appliance, lighting, hardware, and heating specialist retail stores across Victoria participate in the programme by providing information and in-store advice to help guide Victorian consumers to choose energy and water-efficient products at point-of-purchase whether it is in shop or online. Smarter Choice also provides an online energy rating calculator to help consumers and retail staff compare makes, models, and running costs of appliances and products, based on energy rating label and energy consumption.[27]

Actsmart's Home Energy Action Kit

Actsmart's Home Energy Action Kit enables people to assess their energy and water usage at home and then develop a plan to make improvements to save money and the environment. The kit, available to borrow at ACT public libraries, contains:

- A power meter to measure energy consumption and running costs of appliances
- Infrared thermometer to measure fridge, freezer, and hot water temperatures
- A compass to identify the orientation of the home and passive solar heating opportunities
- A stopwatch to measure shower times and tap flow rates
- Instructions on using the equipment and worksheets to calculate home energy efficiency[28]

Star Standard: Energy and Water Efficiency in New Homes

In Victoria, all new homes, home renovations, alterations, and additions need to comply with the 6 Star standard in the National Construction Code. The 6 Star standard applies to the thermal performance of a home, renovation, or addition and includes the installation of either a solar hot water system or a rainwater tank for toilet flushing. The 6 Star energy efficiency rating applies to the home's building envelope: roof, walls, floor, and windows. The 6 Star standard also includes energy efficiency standards for lighting but not plug-in appliances. Minor building works that do not require a building permit are not affected by the regulations. It is estimated that 6 Star homes use 24 percent less energy through heating and cooling compared to 5 Star homes.[29]

New South Wales's Renewable Energy Advocate

The State of NSW has established the Renewable Energy Advocate position to support the state's Renewable Energy Action Plan. The Advocate will work closely with NSW communities and industry and facilitate the development and generation of renewable energy in NSW. The specific tasks of the Advocate include:

- Working across government to facilitate projects
- Attracting investment to NSW by identifying renewable resources and promoting opportunities
- Connecting the banking and finance sector with industry to facilitate underwriting projects
- Reducing and removing complex barriers to investment in renewable energy[30]

11.8   Food: Fiscal Tools

Cities, states and the territory in the Murray-Darling River Basin have implemented a variety of food-related fiscal tools that create interdependencies and synergies between the nexus systems while reducing trade-offs between the systems in the development of a green economy.

Food Source Victoria Fund

Food Source Victoria supports Victorian agri-food businesses in Regional Victoria to work together and with value chain partners to grow exports and create new jobs. The Food Source Victoria fund is a contestable programme to assist the agri-food sector in Regional Victoria increase exports, create new jobs, and develop alliances and value chain partnerships. It consists of two types of funding:

- *Food source planning grants*: Enable agri-food alliances and value chain partnerships based in Regional Victoria to access independent, skilled business advisers to support the development of growth plans

- *Food source growth grants*: Assist alliance and value chain partnerships based in Regional Victoria to implement the improvements recommended in the growth plans[31]

South Australia's Trade Waste Food and Beverage Implementation Grant

South Australia's Trade waste initiative is a two-year programme delivered by the Office of Green Industries SA. The aim of the initiative is to help businesses make improvements to the way trade waste is managed (focussing on quality and quantity), reduce operating costs, and increase productivity by improving the way materials, energy, and water are used. The programme's food and beverage implementation grant offers SA Water trade waste food and beverage customers an opportunity to implement trade waste reduction initiatives at a reduced cost. The grants are available to eligible SA Water food and beverage customers to implement activities that improve trade waste management by reducing volume and/or improving waste quality. Improvements can include new plant and equipment, upgrades, or additions to existing plants and equipment as well as shared infrastructure to reduce waste impacts in a select geographical area. Applicants must exceed or meet 20 percent of at least one of the following waste volume and load-based thresholds and must have completed an approved resource productivity assessment or on-site technology trial to determine a waste solution that is well suited to the business site:

- 10 tonnes biological oxygen demand or suspended solids per year
- 20 tonnes total dissolved solids per year
- 10 million litres per year

A subsidy of up to 50 percent, to a maximum of $300,000, is available to successful applicants for approved implementation activities.[32]

Horticulture Innovation Fund

The Horticulture Innovation Fund is a Victorian Government programme that supports the state's horticulture sector to work together to innovate and adopt new technology and/or processes for economic growth. The objectives of the fund are to:

- Assist the horticulture sector to increase innovation and adopt new technologies that lead to beneficial commercial outcomes for Victoria
- Address regional knowledge gaps relating to productivity and market access
- Strengthen collaboration between the horticulture sector and researchers
- Encourage horticulture businesses across industries to collaborate and share information to increase innovative capacity within the sector
- Promote interest in horticulture research within Victoria

Funding for Eligible Projects

Grants of up to $50,000 are available to eligible applicants to conduct horticulture research and development that will benefit Victoria. As part of the funding, horticulture businesses/organisations and/or associations are required to partner with a research organisation and/or Victorian higher education institution for the conduct of the research and development project. It is expected the projects will be 12 months long. Eligible projects include applied horticulture research and development that is innovative and likely to improve farm performance and/or research that will improve market access and business diversification activities related to the horticulture sector. In addition grant applications must address one of the following themes:

- Innovative production systems, for example, integrated approaches for nutrient and water management
- Managing seasonal variability, for example, use of digital technology to predict and reduce risks of climate change impacts

- Improving quality to meet consumer needs/preferences
- Accessing and maintaining export markets
- Supporting industry transition, for example, innovative re-designing of existing products[33]

11.9   Food: Non-fiscal Tools

Cities, states and the territory in the Murray-Darling River Basin have implemented a variety of food-related non-fiscal tools that create interdependencies and synergies between the nexus systems while reducing trade-offs between the systems in the development of a green economy.

New Horizons: Crop and Livestock Productivity

New Horizons is a Primary Industries and Regions South Australia (PIRSA) initiative to significantly improve broad crop and livestock pasture productivity on South Australia's poorly performing soils. Around 40 percent of South Australia's broadacre farming has soil constraints that could be overcome through the application of new advances in technology, machinery, and soil management. Traditionally, farming has focused on maximising the existing top 10 centimetres of soil. New Horizons instead focuses on modifying the soil profile down to half a metre from the surface to overcome constraints including soil compacting, low fertility, or nutrient-deficient soils and soils with low water-holding capacity. In 2014, the programme began trials across three sites in South Australia with results over the coming years validating potential yield improvements. To provide farmers with the confidence that the techniques will improve yields before they will make the investment in soil modification needed to achieve this increase in productivity, PIRSA over 2015–2016 analysed the economic viability at the farm level and the impact across the state; established new paddock-scale demonstration sites at PIRSA's research centre to look at cost-effective treatment options using locally available organic matter sources; worked with the University of South Australia to design, engineer, and test machinery necessary for

New Horizons practices; and worked with industry and the community to increase knowledge and adoption of the programme. Farmers that do take up New Horizon practices are expected to benefit from:

- Significant increase in soil productivity and profitability
- Improved water use efficiency
- Increased fertility
- Long-term storage of carbon
- Reduced soil erosion risk[34]

City of Sydney: Our City Farm Plan

The City of Sydney has developed a business plan for developing the City Farm into a destination that promotes the role of agriculture and sustainable farming practices. The primary activities of the City Farm include:

- Produce and nursery operations that grow a range of seasonal fruit and vegetables for consumption, generate income through sales at the farmers' market and other outlets, and develop skills and employment opportunities for volunteers.
- Operating a farmers' market that only sells produce directly from growers and enables community engagement and learning through interaction.
- Training and education programmes that increase community interest in where food comes from and provide real opportunities to contribute to food production.
- Volunteering with volunteers acting as ambassadors for the City Farm operations. They will also provide the City Farm with critical human resources required for operations.
- Partnerships with other organisations that align with the visions and values of the City Farm.
- Sponsorship opportunities for corporates, institutions, and organisations to align their activities with the City Farm.
- Composting and waste management with the farm utilising by-products to create compost, establishing an organic waste collection

system to feed the compost operations, and working with surrounding businesses and households to promote composting.

Secondary activities of the City Farm include:

- Animal husbandry programmes that provide a link between rural farming practices and urban communities.
- A farm kitchen that provides a social space for learning programmes and skill development, promotes seasonal produce, and connects local chefs and growers.[35]

Eat Local South Australia

Eat Local SA, an initiative of the Regional Food Industry Association, Food South Australia Inc. and PIRSA, helps consumers locate venues that serve and sell local South Australian food. The initiative supports South Australian food producers and the food service industry that includes local produce in their menus. Participating food venues display Eat Local SA signs as well as feature in the smartphone Eat Local SA itinerary app. South Australian–based food service and food retail businesses are eligible to join Eat Local SA if they meet a series of criteria.

Food services, including restaurants and cafes, must meet a minimum of:

- One main dish with a locally produced core ingredient that names a local producer or regional origin
- Two smaller dishes such as entrees, tapas, or desserts produced from core ingredients from a named local producer or region
- A platter option that includes at least three locally produced core ingredients
- A supplier statement that lists local producers, brands, or regionally sourced core ingredients and names the local producer or regional origin

For food retailers, including farm gates, retailers such as bakeries and greengrocers and so forth, there must be an ongoing commitment to significant promotion of South Australian producers or suppliers with programme participants required to:

- Ensure all products and product descriptions are true to label and origin
- Display the Eat Local SA sign/sticker near the entrance
- Endeavour to use, or offer for sale, the best available quality ingredients and products
- Observe the highest standards of hygiene and food safety
- Support fellow participants in the Eat Local SA programme within the region, or other regions of South Australia[36]

Programme Audits

Food South Australia undertakes regular online and contact-based checks to ensure all participants continue to meet the criteria for participation in the Eat Local SA programme. Participants are required to actively assist in these audit processes on request. If the minimum criteria are found not to be met, Food South Australia will work with the food service or retailer to address any concerns or issues.[37]

11.10   Case Study Summary

Cities, states and the territory in the Murray-Darling River Basin have implemented a variety of initiatives to create interdependencies and synergies between the nexus systems while reducing trade-offs between the systems in the development of a green economy.

General

- The Victorian Government operates ecoMarkets, which is a PES system that pays private landowners for conserving and enhancing the environment. The initiative involves payments for protecting native vegetation on private land, improvements to river and estuary health, and a brokerage system where environmental credits can be traded.

- ACT's Actsmart Schools programme helps schools save energy and water as well as reduce GHG emissions and protect biodiversity. The programme involves providing schools with support to develop environmental plans and conduct water, energy, and waste audits on-site.
- NSW's Sustainability Advantage recognition scheme supports all types of organisations, including non-profits, SMEs, large companies, and government organisations, in identifying and implementing resource efficiency projects. Participating organisations can attend workshops and have one-on-one support on topics including green marketing and behavioural change for sustainability. Participating organisations that show commitment and real improvements are recognised through an awards system.
- ResourceSmart Schools is a Victorian programme that helps schools embed sustainability into all their actions. Schools learn how to manage their water, energy, waste, and biodiversity. Schools can track their progress through an online web portal. Awards are held annually to recognise the most innovative and outstanding sustainability programmes and initiatives taking place in schools.
- NSW's Green Globe Awards recognise organisations of all types, including businesses, public sector agencies, local councils, community organisations as well as individuals, who make the state greener and more innovative.

Water

- The Murray-Darling River Basin operates a water rights trading market based on a cap-and-trade system, with entitlements, which are rights to an ongoing share of the total amount of water available, and allocations, which are the actual amount of water available under water entitlements in a given season, traded among water users.
- South Australia has a statewide pricing system for water services with all non-domestic customers paying a standard rate and domestic customers paying another standard rate for water, with both pricing systems having a fixed and variable rate.

- Queensland has a grant for individuals and communities to reduce water, as well as energy, consumption. Grants can be used for various activities including installing water and energy-efficient pumps and water infrastructure for irrigation systems.
- Victoria's Schools Water Efficiency Program helps schools monitor their water usage. The data is then uploaded onto an interactive web portal enabling schools to use the data as part of their curriculum.
- Sydney's water utility has developed simple benchmarks for businesses to compare their water use with other similar businesses.

Energy

- Victoria offers energy assessment grants for SMEs to increase their energy efficiency as well as grants for industrial associations, business networks, or registered training organisations to develop energy efficiency knowledge and skills in SMEs.
- ACT's business energy and water programme involves an Actsmart assessor conducting an energy assessment, from which the business will be provided a detailed action plan to reduce energy and water usage. Once businesses have implemented the recommendations they can receive rebates for the improvements made.
- Victoria's Smarter Choice retail programme involves retailers providing customers advice on energy- and water-efficient products both online and in-store. The initiative also provides an energy rating calculator and mobile app to help customers and retailers compare makes, models, and running costs of appliances, based on their energy rating label and energy consumption.
- ACT's home energy kit enables people to assess their energy and water usage at home and develop a plan to make improvements to reduce usage and lower bills, with the kit available at public libraries for people to borrow.
- In Victoria, all new homes and renovations will need to comply with the 6 Star standard which typically results in nearly a quarter reduction in energy usage for heating and cooling.
- NSW's Renewable Energy Advocate works with industry to facilitate the development and generation of renewable energy by attracting

investments and promoting renewable energy opportunities as well as connecting the banking and finance sector with industry to facilitate projects.

Food

- Victoria's food source fund assists the agri-food sector in developing alliances and value chain partnerships to enhance productivity, with alliances and partnerships eligible for funding to cover the costs of access to business advisers and funds to implement the improvements recommended by the advisers.
- South Australia's trade waste food and beverage implementation grant enables food and beverage companies to implement projects that reduce waste volume or improve waste quality.
- Victoria's Horticulture Innovation Fund assists the industry in developing new and innovative technologies including systems that involve integrated approaches to nutrient and water management as well as reducing climatic risks.
- South Australia's New Horizons programme enhances crop and livestock productivity on the state's poorly performing soils with the programme's partners working with farmers to increase their knowledge on how to increase soil productivity and improve water use.
- Sydney's city farm plan involves creating a city farm that grows local food, supports a farmers' market, and provides training and education programmes for interested community members.
- South Australia's Eat Local initiative helps consumers locate venues that serve and sell locally produced food with participating food venues displaying Eat Local SA signs. An app has also been developed that displays the participating food venues.

Overall, cities, states and the territory in the Murray-Darling River Basin use a variety of fiscal and non-fiscal tools to create interdependencies and synergies between the nexus systems while reducing trade-offs between the systems in the development of a green economy. These tools are summarised in Table 11.6.

Table 11.6 Murray-Darling River Basin case summary

Tool	Tool type	Policy title	Description	WEF sectors addressed
Fiscal	Market-based instruments and pricing	Trading water rights	The water market is based on a 'cap and trade' system with available water distributed to users via water rights administered by the basin states.	Water
		Water charges	South Australia charges a statewide water price for all domestic customers. All statewide commercial customers pay a fixed and variable charge with the fixed component based on property value.	Water
	Financial incentives	Community sustainability grants: Water	Grants for individuals and community groups to conserve water and use it more efficiently.	Water, energy, food
		Boosting productivity grants	Grants for SMEs to increase their energy efficiency as well as grants for industrial associations, business networks, or registered training organisations to develop energy efficiency knowledge and skills in SMEs.	Energy
		ACT Business energy and water programme	Rebates for ACT businesses upgrading to more water-/energy- efficient technologies.	Water, energy
		Food Source Victoria fund	Fund aims to increase the productivity of the food sector.	Food
		SA Trade waste food and beverage grant	Grant for food and beverage customers to reduce the volume and/or improve the quality of waste.	Food, water
		Horticulture Innovation Fund	Assist horticulturalists to increase productivity.	Food, water
	Payment for ecosystem services	ecoMarkets	Incentives for private landholders to manage their land in ways that conserve and enhance the environment.	Water, food

(continued)

Table 11.6 (continued)

Tool	Tool type	Policy title	Description	WEF sectors addressed
Non-fiscal	Regulations	6 Star standard	All new homes in Victoria need to comply with 6 Star standards of the National Construction Code.	Water, energy
	Public education and skills development	Actsmart Schools	Helps schools save energy, conserve water, and develops skills.	Water, energy
		Schools Water Efficiency Program	Assists schools in monitoring water usage, educates students on technology.	Water
	Public-private partnerships	Renewable Energy Advocate	Works with communities and industry to facilitate the development and generation of renewable energy.	Energy
		New Horizons crop and livestock productivity initiative	South Australia is working with farmers to increase productivity of soils.	Food, water
	Stakeholder participation	City of Sydney farm plan	Plan to promote the development of the inner city farm involving numerous stakeholders.	Food
	Information and awareness-raising	Sydney Water benchmark	Helps businesses compare water use with other similar businesses.	Water
		Smarter Choice retail programme	Provides energy and water efficiency information on consumer goods, appliances.	Energy, water
		Home energy action kit	Helps houses assess energy and water usage and develop a plan to reduce consumption.	Energy, water
	School education	ResourceSmart Schools	Embeds sustainability education into schools.	Water, energy
	Resource mapping	Eat Local SA	Maps food services and retailers that use local produce in their ingredients.	Food
	Awards and public recognition	The Sustainability Advantage recognition scheme	Rewards businesses for their commitment to sustainability and real improvements they make on their sustainability journey.	Water, energy
		ResourceSmart School awards	Recognise outstanding sustainability programmes and initiatives.	Water, energy
		Green Globe Awards	Recognises excellence in environmental sustainability.	Water, energy, food

Notes

1. NSW EPA. 2015. New South Wales State of the Environment. Available: http://www.epa.nsw.gov.au/resources/soe/150817-soe-6-urban-water.pdf.
2. Department of the Environment and Energy, A. G. 2017a. *Climate change impacts in New South Wales* [Online]. Available: https://www.environment.gov.au/climate-change/climate-science/impacts/nsw.
3. Department of the Environment and Energy, A. G. 2017b. Climate change impacts in Victoria.
4. NSW EPA. 2015. New South Wales State of the Environment. Available: http://www.epa.nsw.gov.au/resources/soe/150817-soe-6-urban-water.pdf.
5. Department of Environment, L., Water and Planning. 2017a. *Electricity* [Online]. Available: http://www.delwp.vic.gov.au/energy/electricity.
6. NSW EPA. 2015. New South Wales State of the Environment. Available: http://www.epa.nsw.gov.au/resources/soe/150817-soe-6-urban-water.pdf.
7. New South Wales Government. 2014. Agriculture industry action plan. Taskforce recommendations to government. Available: http://www.dpi.nsw.gov.au/__data/assets/pdf_file/0006/535056/agriculture-industry-action-plan-taskforce_recommendations.pdf.
8. Australian Government Grains Research and Development Corporation. 2014. *Increasing agricultural production by alleviating soil constraints in South Australia* [Online]. Available: https://grdc.com.au/Research-and-Development/GRDC-Update-Papers/2014/02/Increasing-agricultural-production-by-alleviating-soil-constraints-in-South-Australia.
9. Agriculture Victoria. 2016. *Technical report on climate change adaption in agriculture* [Online]. Available: http://agriculture.vic.gov.au/agriculture/innovation-and-research/scientific-reports-and-guides/technical-report-on-climate-change-adaption-in-agriculture.
10. Department of Sustainability and Environment. 2011. *EcoMarkets: valuing our environment,* Melbourne, Department of Sustainability and Environment.
11. Act Government. 2017a. *Actsmart schools* [Online]. Available: http://www.actsmart.act.gov.au/what-can-i-do/schools/actsmart-schools.
12. Office of Environment and Heritage. 2014. Sustainability advantage pathways to recognition. Available: http://www.environment.nsw.gov.au/resources/business/sustainabilityadvantage/140609-SA-recognition-pathways.pdf.

13. Ibid.
14. Sustainability Victoria. 2016a. *Community day a success for 2015 ResourceSmart School of the Year Winters Flat Primary* [Online]. Available: http://www.sustainability.vic.gov.au/news-and-events/media-releases/2016/6/5/community-day-a-success-for-2015-resourcesmart-school-of-the-year-winters-flat-primary.
15. Sustainability Victoria. 2016b. *ResourceSmart schools core module action toolkit* [Online]. Available: http://www.ecorecycle.vic.gov.au/services-and-advice/schools/resources.
16. Office of Environment and Heritage. 2016. Green globe awards 2016: Nomination information guide. Available: http://www.environment.nsw.gov.au/resources/greenglobes/green-globe-nomination-guide-160208.pdf.
17. Murray Darling Basin Authority. 2016a. *Water markets and trade* [Online]. Available: http://www.mdba.gov.au/managing-water/water-markets-and-trade.
18. Murray Darling Basin Authority. 2016b. *Water trading in the Basin* [Online]. Available: http://www.mdba.gov.au/news/water-trading-basin.
19. SA Water. 2017b. *Residential water prices* [Online]. Available: https://www.sawater.com.au/accounts-and-billing/current-water-and-sewerage-rates/residential-water-supply.
20. SA Water. 2017a. *Commercial water prices* [Online]. Available: https://www.sawater.com.au/accounts-and-billing/current-water-and-sewerage-rates/commercial-water-prices.
21. Department of Environment and Heritage Protection. 2016. Community sustainability action grants. Available: https://www.ehp.qld.gov.au/assets/documents/pollution/funding/csa-conservationguidelines-rd1.pdf.
22. Department of Environment, L., Water and Planning. 2017b. *Schools water efficiency program—Data loggers drive savings* [Online]. Available: http://www.depi.vic.gov.au/water/using-water-wisely/education/schools-water-efficiency-program.
23. Sydney Water. 2017. *Benchmarks for water use* [Online]. Available: https://www.sydneywater.com.au/SW/your-business/managing-your-water-use/benchmarks-for-water-use/index.htm.
24. Sustainability Victoria. 2017c. *SV Business—Boosting productivity* [Online]. Available: http://www.sustainability.vic.gov.au/services-and-advice/business/energy-and-materials-efficiency-for-business/boosting-productivity.

25. Sustainability Victoria. 2017a. *Energy efficiency capability grants* [Online]. Available: http://www.sustainability.vic.gov.au/services-and-advice/business/energy-and-materials-efficiency-for-business/boosting-productivity/energy-efficiency-capability-grants.

26. Act Government. 2017b. *Business energy and water program* [Online]. Available: http://www.actsmart.act.gov.au/what-can-i-do/business/business-energy-and-water-program.

27. Sustainability Victoria. 2017b. *Smarter choices* [Online]. Available: http://www.sustainability.vic.gov.au/services-and-advice/households/energy-efficiency/smarter-choice.

28. Act Government. 2017c. *Home energy action kit* [Online]. Available: http://www.actsmart.act.gov.au/what-can-i-do/homes/home-energy-action-kit.

29. Victorian Building Authority. 2017. *6 star standard* [Online]. Available: http://www.vba.vic.gov.au/consumers/6-star-standard.

30. Department of Industry Resources and Energy. 2017. *Renewable energy advocate* [Online]. Available: http://www.resourcesandenergy.nsw.gov.au/investors/renewable-energy/renewable-energy-advocate.

31. Agriculture Victoria. 2015a. Food Source Victoria. Growing the future of food. Fund program guidelines. Available: http://agriculture.vic.gov.au/__data/assets/pdf_file/0003/306039/Food-Source-Victoria-Grant-Guidelines.pdf.

32. Zero Waste SA. 2017. *Food and beverage implementation grants* [Online]. Available: http://www.zerowaste.sa.gov.au/trade-waste/food-and-beverage-implementation-grants.

33. Agriculture Victoria. 2015b. Horticulture innovation fund guidelines and eligibility. Available: http://agriculture.vic.gov.au/agriculture/horticulture/horticulture-innovation-fund.

34. Primary Industries and Regions SA. 2017. *New horizons* [Online]. Available: http://www.pir.sa.gov.au/major_programs/new_horizons.

35. City of Sydney. 2014. Our city farm plan. Available: http://www.cityofsydney.nsw.gov.au/__data/assets/pdf_file/0010/214867/Our-City-Farm-Plan.pdf.

36. Eat Local South Australia. 2017a. *About Eat Local SA* [Online]. Available: http://eatlocalsa.com.au/about/.

37. Eat Local South Australia. 2017b. *Eat Local SA participant criteria* [Online]. Available: http://eatlocalsa.com.au/membership-criteria/.

References

Act Government. 2017a. *Actsmart schools* [Online]. http://www.actsmart.act. gov.au/what-can-i-do/schools/actsmart-schools
———. 2017b. *Business energy and water program* [Online]. http://www. actsmart.act.gov.au/what-can-i-do/business/business-energy-and-water-program
———. 2017c. *Home energy action kit* [Online]. http://www.actsmart.act.gov. au/what-can-i-do/homes/home-energy-action-kit
Agriculture Victoria. 2015a. Food Source Victoria. Growing the future of food. Fund program guidelines. http://agriculture.vic.gov.au/__data/assets/pdf_ file/0003/306039/Food-Source-Victoria-Grant-Guidelines.pdf
———. 2015b. Horticulture innovation fund guidelines and eligibility. http:// agriculture.vic.gov.au/agriculture/horticulture/horticulture-innovation-fund
———. 2016. *Technical report on climate change adaption in agriculture* [Online]. http://agriculture.vic.gov.au/agriculture/innovation-and-research/scientific-reports-and-guides/technical-report-on-climate-change-adaption-in-agricultur
Australian Government Grains Research and Development Corporation. 2014. *Increasing agricultural production by alleviating soil constraints in South Australia* [Online]. https://grdc.com.au/Research-and-Development/GRDC-Update-Papers/2014/02/Increasing-agricultural-production-by-alleviating-soil-constraints-in-South-Australia
Australian Government: Department of the Environment and Energy. 2017a. *Climate change impacts in New South Wales* [Online]. https://www.environ-ment.gov.au/climate-change/climate-science/impacts/nsw
———. 2017b. Climate change impacts in Victoria.
City of Sydney. 2014. Our city farm plan. http://www.cityofsydney.nsw.gov. au/__data/assets/pdf_file/0010/214867/Our-City-Farm-Plan.pdf
Department of Environment and Heritage Protection. 2016. Community sustainability action grants. https://www.ehp.qld.gov.au/assets/documents/pol-lution/funding/csa-conservationguidelines-rd1.pdf
Department of Environment, Land, Water and Planning. 2017a. *Electricity* [Online]. http://www.delwp.vic.gov.au/energy/electricity
———. 2017b. *Schools water efficiency program—Data loggers drive savings* [Online]. http://www.depi.vic.gov.au/water/using-water-wisely/education/ schools-water-efficiency-program

Department of Industry Resources and Energy. 2017. *Renewable energy advocate* [Online]. http://www.resourcesandenergy.nsw.gov.au/investors/renewable-energy/renewable-energy-advocate

Department of Sustainability and Environment. 2011. *EcoMarkets: Valuing our environment*. Melbourne: Department of Sustainability and Environment.

Eat Local South Australia. 2017a. *About Eat Local SA* [Online]. http://eatlocalsa.com.au/about/

———. 2017b. *Eat Local SA Participant Criteria* [Online]. http://eatlocalsa.com.au/membership-criteria/

Murray Darling Basin Authority. 2016a. *Water markets and trade* [Online]. http://www.mdba.gov.au/managing-water/water-markets-and-trade

———. 2016b. *Water trading in the Basin* [Online]. http://www.mdba.gov.au/news/water-trading-basin

New South Wales Government. 2014. Agriculture industry action plan. Taskforce recommendations to government. http://www.dpi.nsw.gov.au/__data/assets/pdf_file/0006/535056/agriculture-industry-action-plan-taskforce_recommendations.pdf

NSW EPA. 2015. New South Wales state of the environment. http://www.epa.nsw.gov.au/resources/soe/150817-soe-6-urban-water.pdf

Office of Environment and Heritage. 2014. Sustainability advantage pathways to recognition. http://www.environment.nsw.gov.au/resources/business/sustainabilityadvantage/140609-SA-recognition-pathways.pdf

———. 2016. Green globe awards 2016: Nomination information guide. http://www.environment.nsw.gov.au/resources/greenglobes/green-globe-nomination-guide-160208.pdf

Primary Industries and Regions SA. 2017. *New horizons* [Online]. http://www.pir.sa.gov.au/major_programs/new_horizons

SA Water. 2017a. *Commercial water prices* [Online]. https://www.sawater.com.au/accounts-and-billing/current-water-and-sewerage-rates/commercial-water-prices

———. 2017b. *Residential water prices* [Online]. https://www.sawater.com.au/accounts-and-billing/current-water-and-sewerage-rates/residential-water-supply

Sustainability Victoria. 2016a. *Community day a success for 2015 ResourceSmart School of the year winters flat primary* [Online]. http://www.sustainability.vic.gov.au/news-and-events/media-releases/2016/6/5/community-day-a-success-for-2015-resourcesmart-school-of-the-year-winters-flat-primary

———. 2016b. *ResourceSmart Schools core module action toolkit* [Online]. http://www.ecorecycle.vic.gov.au/services-and-advice/schools/resources

————. 2017a. *Energy efficiency capability grants* [Online]. http://www.sustainability.vic.gov.au/services-and-advice/business/energy-and-materials-efficiency-for-business/boosting-productivity/energy-efficiency-capability-grants

————. 2017b. *Smarter choices* [Online]. http://www.sustainability.vic.gov.au/services-and-advice/households/energy-efficiency/smarter-choice

————. 2017c. *SV business—Boosting productivity* [Online]. http://www.sustainability.vic.gov.au/services-and-advice/business/energy-and-materials-efficiency-for-business/boosting-productivity

Sydney Water. 2017. *Benchmarks for water use* [Online]. https://www.sydneywater.com.au/SW/your-business/managing-your-water-use/benchmarks-for-water-use/index.htm

Victorian Building Authority. 2017. *6 star standard* [Online]. http://www.vba.vic.gov.au/consumers/6-star-standard

Zero Waste SA. 2017. *Food and beverage implementation grants* [Online]. http://www.zerowaste.sa.gov.au/trade-waste/food-and-beverage-implementation-grants

12

The Green Economy and the Water-Energy-Food Nexus in the Rhine River Basin

Introduction

The Rhine River Basin is characterised by peak discharges in winter and snowmelt and glacier melt in spring, with low water levels in summer and autumn. Around 60 percent of total flow in the lower Rhine originates from sources in the Alps and Southern Germany. During the dry summer and autumn period, this may rise up to 90 percent. Runoff characteristics of the Rhine Basin are governed by two natural buffer systems: the snow buffer and landscape buffer. A significant proportion of winter precipitation accumulates in the Alps in the form of snow and ice, which is gradually released as snowmelt in the spring and early summer. The landscape buffer is based on the landscape acting as a sponge. In its natural state, the landscape buffer system is comprised of extensive natural areas, forests, wetlands, peat lands, and surface water with high water storage capacity. However, large-scale changes in land use has degraded the natural system.

The Convention on the Protection of the Rhine is the basis for international cooperation for the protection of the Rhine. It was signed on 12 April 1999 by representatives of the governments of the five Rhine-bordering countries: France, Germany, Luxembourg, the Netherlands,

© The Author(s) 2018
R.C. Brears, *The Green Economy and the Water-Energy-Food Nexus*,
DOI 10.1057/978-1-137-58365-9_12

and Switzerland and by the European Community. As part of the Convention, these parties are successfully cooperating with Austria, Liechtenstein, the Belgian region of Wallonia, and Italy. Overall, the objectives of the Convention are to ensure:

- The Rhine ecosystem is to be sustainably developed
- Rhine water shall continue to be apt for drinking water production
- The quality of Rhine sediments is to be improved such that dredged material may be deposited without causing environmental harm
- Holistic flood prevention and protection taking into account ecological requirements
- Relief of the North Sea[1]

12.1   Water-Energy-Food Nexus Pressures

The Rhine River Basin is experiencing a variety of water, energy, and food nexus pressures that are detrimental to the development of a green economy as described below through a variety of examples.

Water

Twenty million of the 50 million people who live in the Rhine watershed drink treated Rhine water, which in most cases is produced from river-bank filtrate. The river basin's water is also a key component of the region's EUR 550 billion industrial and chemical industry.[2] The International Commission on the Protection of the Rhine projects that climate change and rising temperatures in the Rhine Basin may lead to changes of precipitation and discharge over the course of the century: during winter, there is likely to be increased precipitation, increased discharge, and early melting of snow/ice/permafrost as well as a shift in the line of snowfall, while in the summer there is likely to be less precipitation but possibly heavy rainfall events followed by dry periods, decreasing discharges, and more periods of low flow. Regarding water temperatures, there will be a strong rise in the numbers of days where Rhine water temperatures exceed

25 °C. The potential impacts on water quality from changes in flow and temperature include:[3]

- Heavy rainfall and flooding
  - More nutrients and pollutants will be carried downstream during short periods of time
  - Pollution may occur from damage to industrial plants, buildings, and so on.
  - Sediments contaminated in the past may be re-mobilised
- Increased precipitation
  - Surface runoff may locally increase pollutant and nutrient loads from non-point-source pollution (e.g. agriculture) and point-source pollution (e.g. combined stormwater system overflows)
- Low flow
  - Discharges of wastewater may cause a rise in concentration of all water quality constituents
  - Salinisation may occur in the Rhine delta[4]

Energy

Almost 80 percent of total electricity generated in the river basin is from nuclear and fossil-fuel-powered plants which depend on the basin's water for cooling purposes. The cooling capacity is likely to decline due to restricted availability of water during long, hot summers and temperature rises. It is projected that over the period 2031–2060 extreme events resulting in power generation reduction will increase by a factor of three. This will impact regions of the Rhine including North Rhine-Westphalia (NRW), which produces around 30 percent of Germany's electricity and 94 percent of the country's coal.[5] With Germany's nuclear phase-out, meaning that nuclear power plants will no longer be used in Germany in 2020, there will be a decrease in net imports from Germany to the Netherlands and by 2025, total electricity exports from the Netherlands will outstrip imports. As a result, production at coal-fired power plants in the Netherlands will increase.[6]

Food

With farmland covering 70 percent of the Rhine Basin, the food industry is a significant economic sector. In the Netherlands, Dutch agricultural exports topped EUR 80 billion in 2014[7] while in the NRW the food industry is worth EUR 30 billion and is the fifth-largest economic sector in the state.[8] The agricultural sector is simultaneously impacted by water surpluses or deficits while being a driver of change in the basin's hydrological regime, for example, in the Netherlands, climate change will likely result in insufficient water supply for agricultural production from the impacts of floods and droughts,[9] while agriculture runoff has led to 76 percent of streams and ditches not meeting water quality standards in one study.[10]

12.2   General Fiscal Tools to Reduce Water-Energy-Food Nexus Pressures

Cities, states, and nations in the Rhine River Basin have implemented a variety of fiscal tools that create interdependencies and synergies between the nexus systems while reducing trade-offs between the systems in the development of a green economy.

Environmental Investment Allowance and Arbitrary Depreciation of Environmental Investments

Using the Environmental Investment Allowance (MIA)/Arbitrary depreciation of environmental investments (Vamil) schemes, Dutch companies can invest in environmentally friendly products or company resources with a fiscal advantage or bring innovative environmentally friendly products onto the market more quickly. Through the MIA, companies can deduct up to 36 percent of the investment costs for an environmentally friendly investment from the fiscal profit on the regular depreciation. With the Vamil, companies may decide themselves when to write off these investment costs. This provides them with an advantage in liquidity and interest. The investments that companies can use the MIA and/or the Vamil for are listed in the Environment List. This includes about 270 investments for which MIA and/or the Vamil can apply. These investments are less damaging to the environment and often go further than legal obligations. For companies who have developed an innovative environmentally friendly product, the development of the market can be stimulated by MIA-Vamil if, following the companies' proposal, their product is included in the Environment List.[11]

12.3 General Non-fiscal Tools to Reduce Water-Energy-Food Nexus Pressures

Cities, states, and nations in the Rhine River Basin have implemented a variety of non-fiscal tools that create interdependencies and synergies between the nexus systems while reducing trade-offs between the systems in the development of a green economy.

Climate Protection Competition: KlimaKita.NRW

In 2015, NRW's EnergyAgency.NRW initiated for the first time a state-wide climate protection competition for children's day centres in order to familiarise young children with climate protection measures. In total 116 day centres across NRW took part in the competition submitting documentation of climate protection projects, with the best works celebrated at the 'Nature and Environment' station at Wuppertal. To facilitate the competition, EnergyAgency.NRW provided assistance and materials for day care employees to establish or expand climate protection projects enabling them to complete the projects successfully.

Green Building Frankfurt Architecture Award

The Green Building Frankfurt award is presented by the City of Frankfurt to builders and planners who make an exceptional contribution to architecture and climate protection. The award recognises innovative, exceptionally well-designed, and sustainable buildings while at the same time drawing broad public attention to award-winning 'green' buildings and to inspire others to do something similar. Every two years energy-efficient houses, school buildings, or office buildings of the Rhine-Main Region can take part in the competition with an expert jury deciding the winners. The focus of the evaluation is on innovative energy concepts, sustainable use of buildings, and an attractive design with all winning projects demonstrating that energy efficiency and sustainability are compatible with expressive design and ease of use.[12] In 2009, eight buildings, ranging from passive houses to energy-efficient office high-rises, were acclaimed as pioneers of sustainable construction in the city.[13]

Guidelines for Economic Building

Frankfurt has developed the *Guidelines for economic building 2013* for all new construction and renovation projects conducted by the city's administration, municipal institutions, city businesses, and all buildings constructed for the City of Frankfurt under public–private partnership contracts. The goal of the guidelines is to minimise annual total costs (sum of capital costs, operational costs, and environmental impact costs) over a building's entire lifecycle (planning, construction, operation, destruction, and disposal) based on a certain quality standard. The level of quality includes:

- Health and comfort for users
- Accessibility for disabled persons
- A local contribution to global climate protection (a 10 percent reduction of carbon emissions in Frankfurt every five years)
- Consideration of climate change already taking place (warmer summers, more intense storms, more frequent flooding)
- Proper designs to ensure users identify with their buildings so they will take better care of the structures
- Protection of high-quality designs and heritage buildings
- Minimisation of material consumption and the primary energy demand of building materials
- Durability and dismantling capacity of structures and components.

Buildings that are deemed in compliance with the guidelines then receive a certificate from the city's Energy Management Department.[14]

Innovation-Orientated Procurement Tool

The Netherlands Enterprise Agency has initiated an innovation-orientated procurement tool, where small businesses that are focusing on innovative solutions to problems including sustainable energy, smart grids, and water management systems can contact the Small Business Innovation Research and submit a tender for a government procurement contract.

Sustainability Clusters in Wallonia

The Walloon government's clustering policy aims to support the development of networks of enterprises in order to promote cooperation and partnership between Walloon enterprises ranging from small to medium enterprises to larger companies. The clustering policy is organised in two structures: business clusters and competitiveness clusters, with business clusters funded to support specific economic areas and competitiveness clusters designed to carry out investment, R&D, or training projects, with both structures designed to promote innovative partnerships whether industrial, commercial, or technological. Clusters that focus on the sustainable use of natural resources include:

- *CAP2020*: This is a cluster dedicated to the building industry, bringing together contractors, architects, manufacturers, and suppliers of materials and services who are committed to developing and promoting sustainable building in Wallonia.
- *Green buildings cluster*: The cluster brings together enterprises, architects, study centres, and professionals who incorporate the principles of green building in their activities.
- *TWEED cluster*: TWEED (Technology of Wallonia Energy, Environment and Sustainable Development) is a cluster that includes over a hundred enterprises in the sector of renewable energy and energy efficiency in the tertiary and industry sectors.[15,16,17,18]

Demonstration Project: 100 Climate Protection Housing Estates

The 100 Climate Protection Housing Estates project of EnergyAgency. NRW shows how homes and living will develop in the future. To date 64 estates have been awarded the status of 'Climate Protection Housing Estate' in NRW, 18 of which are already complete, 28 are undergoing construction, and 18 are still in the planning stages. Currently 3000 people live in 1300 houses and apartments in the climate protection estates. The residential units of the most recent estate inaugurated emit only 3.5 tonnes of CO_2 per annum: 90 percent less than emissions from conventional buildings.

City Deals

City Deals are the means of achieving the objectives of Agenda Stad (Agenda City): an initiative by the Dutch government to promote the competitive power of Dutch cities and to keep the Netherlands in the global top by strengthening growth, innovation, and quality of life in the Dutch cities. In City Deals concrete cooperation agreements between cities, national authorities, local authorities, private enterprises, and NGOs are anchored. Those deals must lead to innovative solutions for societal issues or include measures to strengthen the economic ecosystems of the urban region. The aim is to interlink ambitious and powerful players in the urban network. City Deals are thus intended to create new forms of cooperation, through which urban challenges are addressed in an efficient manner. The cities themselves decide, in consultation with the ministries concerned, the problem that is addressed in each City Deal.

Green Deals

In the Netherlands, Green Deals are agreements between the central government and other parties: private companies, civil society organisations, and other authorities (provinces, municipalities, water boards). The Green Deal helps to carry out sustainable plans, for example, for energy, climate, water resources, biodiversity, mobility, bio-based economy, construction, and food. The role of the national government in Green Deals is to help in eliminating bottlenecks in sustainable plans. It can do so in several ways:

• Working on adaptations in laws and regulations, to reduce the administrative burden on businesses
• Acting as a mediator, for example, to bring together organisations or get difficult negotiations in motion again
• Helping businesses to develop new markets for sustainable technology, for example, by helping companies to enter foreign markets ('green trade missions')

12.4   Water: Fiscal Tools

Cities, states, and nations in the Rhine River Basin have implemented a variety of water-related fiscal tools that create interdependencies and synergies between the nexus systems while reducing trade-offs between the systems in the development of a green economy.

Pricing Water to Encourage Efficiency and Protect Water Quality

In the Netherlands, the efficiency of industrial water use has increased significantly over the past decades. Between 1976 and 2012, the industrial use of groundwater has fallen by 70 percent. The industrial use of water overall has fallen by 20 percent between 1976 and 2012. There are a number of incentives that played a role. First, the use and pollution of water is charged: per cubic metre of drinking water or groundwater, and per unit of pollution for wastewater discharge. As a result of these charges companies will try to reduce their use of drinking water or groundwater and ensure that they minimise the amount of discharged pollution units.

In addition, water savings often lead to lower energy costs, since process water is often cooled or heated.

EIA: Tax Deductions for Investments by Water Companies in Energy-Saving Measures

Water companies can use the Energy Investment Allowance (EIA) to invest in energy-efficient technology and durable energy under favourable fiscal conditions. Companies can deduct 58 percent of the investment costs from the fiscal profits, on top of their usual depreciation. As a result, companies pay less income tax or company tax. On average, the EIA, commissioned by the Ministry of Economic Affairs, results in a 14 percent tax advantage. Alongside this tax advantage, energy-efficient investments also ensure a lower energy bill. The available options for water companies are listed in the annual Energy List of about 150 EIA-eligible investments.[19]

12.5   Water: Non-fiscal Tools

Cities, states, and nations in the Rhine River Basin have implemented a variety of water-related non-fiscal tools that create interdependencies and synergies between the nexus systems while reducing trade-offs between the systems in the development of a green economy.

Water Knowledge Programme: Lumbricus

The Netherlands' new knowledge programme Lumbricus involves water boards, research institutes, and SMEs joining forces to design and investigate climate-robust solutions to the current soil and water issues on the higher sandy soils in the Netherlands. Climate change causes problems such as periodic flooding and periodic droughts, which are also linked to an increase of problems such as eutrophication, inadequate freshwater availability, deterioration of water quality, degradation of the natural values, and decreased agricultural potential.

Water Quality Protection: Activities Decree

To protect the environment against negative impacts from agriculture the Netherlands' Activities Decree contains environmental regulations, especially for businesses. All companies in the Netherlands have to comply with the Activities Decree, unless they have no physical device present in the environment. The Activities Decree contains rules by type of environmentally harmful activities and by type of environmental impact. The Activities Decree has different rules for different types of devices. The rules determine which standards companies must meet when it comes to water use. The legislation on nutrients is similar. This legislation sets limits on the amount of nutrients a farmer may apply. If a farmer produces more manure than they can use on their soil, manure treatment is mandatory. This regulation reduces the pollution of groundwater with nutrients.

Delta Plan for Agricultural Water Management

The Delta Plan for Agricultural Water Management (DAW) describes how agriculture and horticulture can help in solving the water challenges, coupled with the strengthening of the agricultural and horticultural sectors. In order to achieve this goal, an attempt is made to:

- In 2021, solve 80 percent of the remaining water quality problems in a motivating and stimulating way, and in 2027, 100 percent.
- In 2021, to make agricultural water supply sustainable by reducing water use at farm level, apply water conservation at regional level, and a smarter distribution and buffering at national level, in line with the delta decision taken in 2014.
- By means of tailor-made approaches, new spatial tools, and innovative techniques, increase the agricultural production potential on a regional level by 2 percent per year.

The DAW intends to have both public and private parties collaborate on developing innovative solutions to water problems. Also, best practices are collected in order to be consulted by other parties who want to apply the same measure. A comprehensive cooperation in the water sector is thus central to the DAW.

The Water Nexus Project: Salt Water Where Possible, Freshwater Where Necessary

Water Nexus is a new joint Netherlands Organisation for Scientific Research (NWO)-STW research programme which will run until 2020, and is supported by 25 partners from multinational and small- and middle-sized companies, consultancy firms, research institutes, water boards, and the Ministry of Infrastructure and Environment. Researchers work together to develop new solutions for water supply in freshwater-scarce coastal regions in the Netherlands and abroad. The project focuses on strategies to balance water supply and water demand between industrial and agricultural systems and contains both saline and fresh water options. The research project includes water technology and green technologies such as wetlands. The emphasis will be on industry and agri-/horticulture where large volumes of water are produced and used. Technologies and management strategies will be developed that enable climate change adaptation by exchange of water between industry, agriculture, urban, and natural systems; the use of brackish water as a resource; and the inclusion of green infrastructure for water storage and treatment.[20]

Partners for Water

The Netherlands has extensive knowledge and experience in the field of water and water management and has built up an international reputation. The international growing demand for water knowledge offers opportunities for the Dutch water sector. The Partners for Water programme is committed to support the Dutch water sector in capitalising on these opportunities. One of the main initiatives included is the water safety and water security in urbanising deltas grant programme. This grant scheme provides opportunities for businesses, knowledge institutions, and NGOs to test innovative technologies, methodologies, or prototypes in water management and to demonstrate or investigate their feasibility in urban deltas abroad. The subsidy covers the following topics:

- Drinking water and sanitation
- Water governance
- Sustainable development of waterways and harbours (on shore)
- Climate (water) security
- Food and ecosystems
- Water and energy

12.6   Energy: Fiscal Tools

Cities, states, and nations in the Rhine River Basin have implemented a variety of energy-related fiscal tools that create interdependencies and synergies between the nexus systems while reducing trade-offs between the systems in the development of a green economy.

Energy-Saving Cash Bonuses

Frankfurt is the first city in Germany to reward electricity savings with cash bonuses. Customers who reduce their electricity consumption by at least 10 percent within a year receive a bonus of EUR 20 from the city plus 10 cents for every additional kilowatt-hour saved. On average participants have saved 24 percent electricity in one year and received a bonus of around EUR 70 for their savings.[21]

Energy-Saving Support for SMEs

The City of Frankfurt helps small- to medium-sized enterprises, as well as clubs and community centres, accurately analyse their energy consumption and find ways of reducing their energy usage and electricity bills. The programme involves the organisation contacting KfW who then inspects the premises and writes a report on the organisation's energy-saving potential. The report determines investment measures that will lead to reduced energy usage with savings clearly calculated. After implementing the recommended energy-saving measures, customers are eligible to get 10 cents back for every kilowatt-hour saved. To help SMEs implement energy-saving measures the city funds up to a maximum of 30 percent of the total investment with total funding limited to EUR 50,000 per year per company with funding paid upon proof of implementation.[22]

Energy Investment Allowance

The Netherlands Enterprise Agency's EIA enables companies to invest in energy-saving equipment and sustainable energy with tax benefits. As part of the scheme companies can deduct 41.5 percent of the investment cost from taxable profits, resulting in lower income and corporate taxes. Investments in eligible assets are listed in the Energy List, which is updated annually. In particular, the Energy List contains examples of investments in energy-efficient appliances, systems, and sustainable technologies. In addition, companies can request tax deductions for customised investments that meet the energy performance criteria stated in the Energy List.

12.7   Energy: Non-fiscal Tools

Cities, states, and nations in the Rhine River Basin have implemented a variety of energy-related non-fiscal tools that create interdependencies and synergies between the nexus systems while reducing trade-offs between the systems in the development of a green economy.

EnergyRegion.NRW Cluster

The NRW state government has established the 'EnergyRegion.NRW' cluster to achieve its energy and climate policy. The cluster lies within the framework of the state's cluster policy of creating a favourable environment for innovations that will boost the competitiveness of industry and stimulate growth and employment. The main aim of EnergyRegion. NRW is to accelerate processes of innovation in the energy economy and scale them up to market readiness both domestically and internationally. This is achieved by:

- Providing a cooperative network for small- to medium-sized companies, large companies, multinationals, research institutes, and political decision-makers
- Offering cross-company projects that link partners and provide neutral information from one single-source
- Enabling large companies to identify small, innovative companies in the region
- Providing up-to-date information from authorities and political decision-makers to industrial companies

The EnergyRegion.NRW cluster comprises eight subject areas including biomass, fuel cells and hydrogen, energy-efficient and solar construction, geothermal energy, fuels and engines of the future, power plant technologies, photovoltaics, and wind power, each of which comes under networks.

Municipal-Scale Information on Geothermal Energy Potential

On behalf of NRW's State Environmental Agency, the International Geothermal Centre and the Department of Geodesy at Bochum University of Applied Sciences have determined the potential for near-surface geothermal energy in the state. Unlike the existing potential map produced by NRW's Geological Surface, this map not only projects the geothermal yield of a location being investigated but has also been set against the specific heat requirement of the buildings of a specific site. This enables the viewer to see at the municipal level the percentage of actual heat requirement that could be met by heat pumps in combination with geothermal probes that reach a maximum depth of 100 metres.

Energy(-Savvy) Village

The Energy(-savvy) Village is a study initiated by the South Westphalia agency, with funding from the NRW Climate Protection Ministry (progress.NRW), where village communities are supported and coached on how to integrate village developments with climate protection measures. The programme involves village cars, communal energy concepts, education projects, excursions to 'Good Examples', and holding of brainstorming workshops. The objectives of the programme are to drive the mobility and energy revolution and raise awareness of potential climate change measures that can be implemented locally. The results of the study will form a guideline for other villages in the future to learn from in the transition to Energy(-savvy) Villages.

Klimakidz Project: Renewable Energy Education

The Klimakidz project of EnergyAgency.NRW introduces young people to the basic themes of renewable energy. The project involves students learning about renewable energy in a fun, informative way with practical in-class experiments. The project is targeted at fifth and sixth graders of all secondary schools and is available year-round to schools free of charge.

The Netherlands' Agreement on Energy for Sustainable Growth

In 2013, the Netherlands' Agreement on Energy for Sustainable Growth was signed by 47 organisations including local governments, employers' associations and unions, environmental organisations, financial institutions, NGOs, and other stakeholders. The negotiations were held over an eight-month period with independent chairs leading discussions on the four major themes of the Agreement: energy saving, renewable energy, innovation, and the transport sector. During the negotiations, independent research institutes, the Energy Research Centre of the Netherlands, and the Netherlands Environmental Assessment Agency played a key role by estimating the effects of the proposed actions, giving confidence to participating organisations that proposed actions could meet national and European Union objectives. Signatories to the Agreement share responsibility and commitment to achieve four objectives:

1. An average energy efficiency saving of 1.5 percent per year (adding up to a reduction of 100 PJ by 2020)
2. 14 percent of renewable energy in the Netherlands' total consumption of energy by 2020
3. And 16 percent by 2023
4. Creating at least 15,000 additional jobs by 2020

Implementing the Agreement

To meet the four main objectives the Agreement consists of 12 pillars, each with its own actions. Each action has been assigned to a representative of one of the signatory parties to the Agreement while each pillar has one or two coordinators, who are representatives of organisations that have signed the Agreement. A Standing Committee oversees progress as well as organises conferences to exchange and deepen relevant knowledge and expertise. Transparent monitoring of progress is fundamental to the tasks of the Standing Committee, which is facilitated by several tools that include:

* A monitor (dashboard) to keep track of the progress of actions, results, and expected effects
* An annual progress report of the Agreement
* Annual analysis in the Dutch National Energy Report
* An evaluation in 2016 of the Agreement and the Committee's operations[23]

Guidelines for Incorporating Energy Savings in Historical Building Renovations

In Frankfurt, significant energy savings can be made from renovating existing buildings: Around 15,000 buildings, the majority of which are built in the late nineteenth century and early 1900s, need to be brought into line with twenty-first-century energy efficiency standards. Frankfurt's Energy Agency has published the brochure *Guidelines for the energy-saving renovation of Gründerzeit buildings* to show that conservation of historical buildings and climate protection are not mutually exclusive. The brochure notes that even in buildings with façades and roofs that are intricately designed and worth conserving heating demand can be reduced by 50 percent without infringing the city's regulations governing the protection of historic buildings.[24]

Public–Private Partnership in the Biobased Delta

Under the name Biobased Delta entrepreneurs, institutes, and government agencies in South-western Netherlands are working together to develop a bio-based economy. The Biobased Delta in particular is investigating and developing alternative fuels to replace or complement fossil fuels. One area of experimentation is the use of algae in purifying water rather than relying on energy, as well as chemicals, to treat wastewater. A large project at Yara Sluiskil uses algae to purify water from a fertiliser factory. Yara uses around 600,000 litres of water per day to make its products with the wastewater, containing residues including nitrogen, CO_2, and other minerals, discharged into the Ghent-Terneuzen Canal. Research is underway to establish whether the purification of wastewater with algae in an open reservoir is economically and technically feasible. The pilot study was launched in March 2014 with EUR 1 million invested. In 2016 an evaluation was conducted on whether the process has potential to attract further investment from businesses.[25]

12.8   Food: Non-fiscal Tools

Cities, states, and nations in the Rhine River Basin have implemented a variety of food-related non-fiscal tools that create interdependencies and synergies between the nexus systems while reducing trade-offs between the systems in the development of a green economy.

Netherlands Sustainable Food Alliance

The Sustainable Food Alliance and the Ministry of Economic Affairs have drawn up a joint agenda that by 2020 the Netherlands' agri-food chain will have undergone a complete sustainable transition. Specifically, by 2020 all food available on the Dutch market will be produced and distributed significantly more sustainably and the amount of food waste will be significantly lower than in 2013. To achieve this goal, the Sustainable Food Alliance undertook a range of 'best effort' commitments for the period 2013–2016 that focused on four areas of action:

1. Increasing the sustainability of the food chain in a broad sense, with extra attention paid on:
   1. Making the meat chain more sustainable
   2. Reducing food waste and optimising residual waste streams
   3. Improving the transparency of and communication about the transition to a more sustainable food production and distribution system

To facilitate these actions, the Ministry of Economic Affairs will conduct activities including:

- Ensuring knowledge generated by innovative pilots and best practices are shared with various stakeholders
- Supporting the market development of the premium segment through trade missions
- Removing obstacles in existing legislation and regulations
- Informing customers on the environmental benefits of sustainable food
- Developing means to making transparent food chains and production processes transparent for consumers

The Sustainable Dairy Chain

In 2014, the Dutch Federation of Agriculture and Horticulture (LTO), representing around 70 percent of the 18,000 Dutch dairy farmers, and the Dutch Dairy Organisation (NZO), whose membership comprises 13 dairy companies that process 98 percent of all milk made in the Netherlands, formed ZuivelNL to strengthen the Dutch dairy chain in a way that respects the environment and society. For the government ZuivelNL is the organisation for the Dutch dairy supply chain that forms a neutral point of contact of the dairy industry. ZuivelNL has been approved by the Dutch government as an inter-branch organisation, in accordance with European law, in order that agreements in areas of substantial public interest can be declared universally binding.

The Sustainable Dairy Chain (Duurzame Zuivelketen) is part of ZuivelNL and aims to create a dairy sector that is future-proof and responsible. In particular, a sector where work is satisfying and safe, where one can earn a good living, which produces high-quality food, which respects animals and the environment is appreciated. To achieve a future-proof and responsible dairy sector, the Sustainable Dairy Chain has established four visions to achieve by 2020:

1. Development towards climate neutrality
2. Continuous improvements in livestock health and welfare
3. Preservation of grazing
4. Protection of biodiversity and the environment

Regarding the visions of development towards climate neutrality and protection of biodiversity and the environment, the Sustainable Dairy Chain has developed a list of goals to achieve by 2020 and indicators to measure progress (Table 12.1).[26]

Implementation: Energy Scan

FrieslandCampina has developed an energy scan for members of the Sustainable Dairy Chain to compare their energy use with other companies and dairy farmers of comparable size. As part of the initiative, dairy farmers complete an Internet-based questionnaire on energy consumption, equipment, type of milking parlour, and lighting: For the first time

Table 12.1 Sustainable Dairy Chain visions, goals, and indicators

Vision	Topic	Goal	Indicators
Development towards climate neutrality	Greenhouse gases	20 percent reduction of greenhouse gases in 2020 compared to 1990 and climate neutral growth compared to 2011	Dairy chain's CO_2-equivalent Milk's CO_2-e.per kg
	Sustainable energy	16 percent sustainable energy production by 2020	Sustainable energy (%) Share of sustainable energy in energy consumption (%)
	Energy efficiency	Energy efficiency improves on average 2 percent per year between 2005 and 2020	Primary fuel consumption per kg of milk
Protection of biodiversity and the environment	Responsible soy	100 percent use of responsible soy as of 2015	Share of responsibly purchased soy feed (%)
	Nutrients	Total farming phosphorous production stays below European limit (173 million kg) Keep dairy farming phosphorous production at 2002 level (84.9 million kg) 5 kton reduction in ammonia emissions compared to 2011	Phosphorous separation in manure from farming and dairy herds (million kg) Amount of ammonia from animal manure from dairy herds (million kg NH_3)
	Biodiversity	No net reduction in biodiversity. By 2017 a monitoring system is developed and concrete goal set	Composite score of stress on biodiversity Share of farms that apply environmental management and/or are members of ANV (%)

a complete inventory is required, after which only updates are required. The information entered is then compared to the anonymous energy scan database, which is updated regularly. From the database, dairy farmers receive an energy scan report based on their answers to the questionnaire with the report stating how many kilowatt-hours per 1000 litres of milk are used compared to the total group of dairy farmers as well as operators with a similar type of milking parlour. Simultaneously, dairy farmers receive tips for energy saving including on lighting. For farmers that wish to make additional energy savings, workshops are also organised by FrieslandCampina.[27]

Sustainable Soy

As of 1 January 2015, the Dutch dairy sector, which buys around 300,000 tonnes of soy per annum, only purchase soy that is Roundtable on Responsible Soy (RTRS) certified. Soy that conforms to the RTRS standards protects habitats and respects human rights in its production. In addition to grass, cows also eat concentrated feed that often contains soy. However, worldwide soybean production often leads to deforestation, encroachment into special habits, and large-scale use of pesticides.[28]

Sustainable Biomass

The Netherlands does not produce enough biomass to meet the needs of its energy and chemical sector and therefore needs to import biomass. To increase the sustainability of biomass production, particularly in developing countries, the Netherlands Programme Sustainable Biomass (NPSB) was developed to gain experience in the production and certification of sustainable biomass. The NPSB programme ran from 2009 to 2013, during which the programme consisted of projects from the Global Sustainable Biomass tenders (DBM) and from the Sustainable Biomass Import tenders (the DBI projects) with projects—funded by the Dutch Ministry of Economic Affairs and Ministry of Foreign Affairs—designed to complete knowledge gaps with supplementary research. Specifically, the DBM projects aimed to stimulate, support, and facilitate the promotion of sustainable production, processing, and importing of biomass produced in developing countries, while the DBI projects had a similar aim but focused on the production of biomass for export to the Netherlands.[29]

The Wagralim Cluster

The Wagralim cluster is a competitiveness cluster for the agro-industry in Wallonia. The cluster brings together companies, training centres, and public and private research centres into competitive projects with the common objective of participating in high value-added projects. Wagralim offers various services to companies including support on R&D projects, strategic analyses of sectors, training, and assistance in international development with the overall aim being to make members more competitive and innovative both regionally and internationally by increasing activity and employment, and developing a culture of innovation. The Wagralim strategy focuses on four priority fields of:

- Food/nutritional quality: Relating to the design, development, and assessment of products and ingredients with proven health benefits
- Industrial efficiency: Aims to improve manufacturing processes and methods in food preservation
- Packaging: Creating tomorrow's packaging that uses renewable raw materials
- Development of sustainable agro-industry networks: Aims to optimise production processes and by-products to enhance sustainability[30]

Creation of the Nutrient Platform

The Nutrient Platform is a cross-sectoral network of Dutch organisations that believe in a pragmatic approach towards nutrient scarcity. Thirty-six

Dutch parties from the water sector, agriculture, waste sector, and chemical industry, from businesses, knowledge institutes, NGOs, and national ministries, have joined forces to close nutrient cycles. The Nutrient Platform aims to accelerate the transition towards sustainable nutrient management by creating a market for recycled nutrients. Above all, the Nutrient Platform aims to turn the surplus of phosphorus in the Netherlands into an opportunity. This surplus, mainly the result of the Netherlands' intensive livestock system, causes environmental problems. By recovering phosphorus from 'waste' streams and turning it into valuable new products, not only is the environment improved but also the phosphorus cycle is closed.[31]

12.9   Case Study Summary

Cities, states, and nations in the Rhine River Basin have implemented a variety of initiatives to create interdependencies and synergies between the nexus systems while reducing trade-offs between the systems in the development of a green economy.

General

- Dutch companies can use the MIA and Vamil schemes to invest in environmentally friendly products or company resources in return for fiscal benefits or bringing the product or service to the market.
- NRW has a statewide climate protection competition for children to familiarise them with climate protection measures.
- Frankfurt's green building architecture award is given to builders and planners who make exceptional contributions to architecture and climate protection.
- Frankfurt has created *Guidelines for economic building*, which aims to reduce the environmental impact of all new construction and renovation projects constructed under public–private partnerships.

- The Netherlands' innovation-orientated procurement tool enables small businesses focusing on innovative solutions to resource efficiency to submit tenders for government procurement contracts.
- Wallonia has developed clusters of enterprises ranging from SMEs to large companies that focus on various sustainability and resource efficiency issues.
- NRW's 100 climate protection housing estate project shows how houses of the future can emit low amounts of GHG emissions.
- The Netherlands' City Deals involve the national government working with various stakeholders to initiate innovative solutions to societal issues.
- The Netherlands' Green Deals are between the national government and various stakeholders to facilitate the development of sustainability plans with the government reducing administrative and regulatory burdens on businesses.

Water

- The Netherlands charges industrial water uses for groundwater usage and for pollution discharged into waterways.
- The Netherlands' EIA enables water companies to invest in energy-efficient technologies with lower tax rates on income or company tax.
- The Netherlands' Lumbricus programme involves stakeholders joining forces to design and investigate climate-robust solutions to water and soil issues.
- All companies in the Netherlands are bound by the Activities Decree, which contains rules on environmentally harmful activities and by type of environmental impact and determines which standards a company must meet when it comes to water use.
- The Netherlands' Delta Plan for Agricultural Management sets out how agriculture and horticulture can solve water challenges, including initiating tailor-made approaches at the regional level.

- The Water Nexus Project involves multiple stakeholders with support of the Dutch government working together to develop new solutions for water supply in freshwater-scarce coastal regions in the Netherlands and abroad.
- The Netherlands' Partners for Water programme involves the exporting of Dutch water technology and know-how with grants provided to Dutch companies to demonstrate innovative technologies abroad.

Energy

- Frankfurt provides cash bonuses to customers who reduce their electricity consumption by a certain amount each year.
- Frankfurt helps SMEs as well as non-profits analyse their energy usage and find ways of reducing consumption and lowering their electricity bills.
- The Netherlands' EIA provides tax rebates for companies that invest in energy-efficient equipment and sustainable energy.
- NRW's energy region cluster aims to accelerate the state's energy economy and scale up technologies for market readiness both nationally and internationally.
- NRW has mapped the potential for geothermal energy and determines at the municipal level the percentage of actual heat that could be met geothermally.
- NRW's Energy(-savvy) Village project involves village communities receiving expert advice on how to integrate climate protection measures including communal energy systems.
- NRW's Klimakidz project involves students learning about renewable energy in a fun way with practical in-class experiments.
- The Netherlands' Agreement on energy for sustainable growth is a multi-stakeholder agreement across the public and private sectors to increase energy efficiency and renewable energy and create additional jobs in the process.
- Frankfurt has created guidelines for how renovations to existing historical buildings can incorporate energy efficiency measures.

- Government agencies as well as entrepreneurs and institutes in the South-western Netherlands are working together to form a bio-based economy with the partnership developing alternative fuels.

Food

- The Netherlands' Sustainable food alliance and the government have drawn up an agenda for the agri-food industry to undergo a complete sustainable transition, with the government providing an array of support services including sharing best practices, and informing consumers on the environmental benefits of sustainable food.
- The Dutch government has approved the Sustainable Dairy Chain Initiative's ZuivelNL organisation as being an inter-branch organisation, ensuring that agreements in areas of substantial public interest can be declared universally binding. The organisation has developed a series of visions for the industry to achieve by 2020 including reducing GHG emissions, increasing energy efficiency, and reducing nutrient use on farms.
- The Netherlands conducted a wide study to support and facilitate the production and processing of sustainable biomass in developing countries.
- Wallonia's Wagralim cluster aims to increase food and nutritional quality, increase industrial food production efficiency, enhance the sustainability of packaging, and optimise production process and by-products to enhance sustainability.
- The Netherlands' nutrient platform involves a wide range of stakeholders from the public and private sectors working together to create a market for recycled nutrients recovered from waste streams.

Overall, cities, states, and nations in the Rhine River Basin use a variety of fiscal and non-fiscal tools to create interdependencies and synergies between the nexus systems while reducing trade-offs between the systems in the development of a green economy. These tools are summarised in Table 12.2.

Table 12.2 Rhine River Basin case summary

Tool	Tool type	Policy title	Description	WEF sectors addressed
Fiscal	Market-based instruments and pricing	Pricing of water and wastewater	The Netherlands prices water and wastewater to encourage industry to use water efficiently and reduce pollution from wastewater.	Water, energy
	Environmental taxes	Energy Investment Allowance/Arbitrary depreciation of environmental investments (Vamil) schemes	Water and other types of companies can use the allowance to invest in energy efficiency, with the allowance in the form of a tax deduction, while Vamil allows Dutch companies to invest in environmentally friendly products or bring innovative environmentally friendly products into the market more quickly with tax benefits.	Water, energy, food
	Financial incentives	Partners for Water	Grants scheme for businesses, institutions, and NGOs to test innovative water technologies and their feasibility internationally.	Water, energy, food
		Energy-saving bonuses	Electricity customers are rewarded for saving electricity with cash bonuses.	Energy
		Energy-saving support for SMEs	Following an audit, funding is available for SMEs to implement energy-saving recommendations.	Energy
Non-fiscal	Regulations	Activities Decree	The Netherlands' Activities Decree contains rules that must be met by companies when using water as well as the amount of nutrients farms may apply.	Water, food
	Public-private partnerships	*Guidelines for economic building*	The guidelines for new construction and renovation projects conducted under public–private contracts aim to minimise total costs including environmental impact costs.	Water, energy

(continued)

Table 12.2 (continued)

Tool	Tool type	Policy title	Description	WEF sectors addressed
		City Deals	City Deals aim to interlink the public and private sector, as well as NGOs, in developing innovative solutions for societal issues.	Water, energy, food
		Green Deals	Green Deals are agreements between the public and private sector, including NGOs, to carry out sustainability plans, with the government facilitating the process.	Water, energy, food
		Biobased Delta	Entrepreneurs, institutes, and government agencies are working together to develop alternative fuels, as well as develop algae to treat wastewater.	Energy, water
		Delta Plan for Agricultural Water Management	Public and private parties collaborate on the development of innovative solutions to water problems in the agricultural sector as well as sharing of best practices.	Water, food
		Water knowledge programme: Lumbricus	The Netherlands' water knowledge programme involves water boards, research institutes, and SMEs joining together to research water-related issues.	Water
	Stakeholder participation	Sustainable Biomass	With government funding support, projects were undertaken by various stakeholders to increase the sustainability of biomass exported to the Netherlands from developing countries.	Food, energy
		Nutrient platform	A cross-sector of the Netherlands, including the government, private sector, institutes, and NGOs, is joining forces to close nutrient cycles.	Food, water

(continued)

Table 12.2 (continued)

Tool	Tool type	Policy title	Description	WEF sectors addressed
	Information and awareness-raising	Guidelines for incorporating energy savings in historical buildings	The guidelines show that conservation of historical buildings and climate protection are not mutually exclusive.	Energy
	School education	Klimakidz project	The project introduces young people to the basics of renewable energy through in-class experiments.	Energy
	Clustering policies	Sustainability clusters in Wallonia	The industrial clusters focus on green buildings and sustainable energy.	Water, energy
		Water Nexus Project	Government agency researchers as well as multinational and SMEs, research institutes, water boards, and consultancy firms work together to develop new solutions for water supply in freshwater-scarce coastal regions in the Netherlands.	Water, food
		Energy cluster	NRW has established clusters in the areas of biomass, fuel cells and hydrogen, solar, wind, and so on.	Energy
		Wagralim cluster	The agro-food cluster aims to increase food quality and industrial efficiency, and optimise the use of by-products.	Food
	Resource mapping	Geothermal energy potential	The map provides municipal-level information on the percentage of actual heat required that could be met by geothermal.	Energy
	Public procurement	Innovation-orientated procurement tool	Small businesses focusing on innovative solutions can submit tenders for government procurement contracts.	Water, energy, food

(continued)

Table 12.2 (continued)

Tool	Tool type	Policy title	Description	WEF sectors addressed
	Demonstration projects	100 Climate Protection Housing Estates	The demonstration project shows how homes and living will develop in the future with low-carbon living.	Water, energy
		Energy(-savvy) Village	Village communities are supported and coached on climate protection initiatives with the goal of becoming energy-savvy.	Energy
	Awards and public recognition	Climate protection competition	The climate protection competition involved children day centres initiating climate protection measures with the best works celebrated.	Water, energy
		Green Building Frankfurt architecture award	Award to builders and planners who make exceptional contributions to architecture and climate protection.	Water, energy
	Voluntary agreements	Agreement on Energy for Sustainable Growth	Nearly 50 organisations from the public and private sector as well as unions signed an agreement to increase energy efficiency and renewable energy and create new jobs.	Energy
		Netherlands Sustainable Food Alliance	The government and the private sector have drawn up an agenda for the agri-food sector to undergo a complete sustainable transition.	Food
		Sustainable Dairy Chain	The Sustainable Dairy Chain, part of ZuivelNL which is approved by the Dutch Government to make binding agreements for all members, aims to work towards climate neutrality and environmental protection.	Food, water, energy

Notes

1. ICPR. 2017a. *Convention on the protection of the Rhine* [Online]. Available: http://www.iksr.org/en/international-cooperation/legal-basis/convention/index.html.
2. ICPR. 2017b. *Industry* [Online]. Available: http://www.iksr.org/en/uses/industry/index.html.
3. ICPR. 2015. Strategy for the IRBD Rhine for adapting to climate change. Available: http://www.iksr.org/fileadmin/user_upload/Dokumente_en/Reports/219_en.pdf.
4. Ibid.
5. NRW Invest. 2011. Energy in North Rhine-Westphalia. Available: http://www.energieregion.nrw.de/_database/_data/datainfopool/Energy_in_North_Rhine-Westphalia_Facts_Figures.pdf.
6. PBL Netherlands Environmental Assessment Agency. 2016. National energy outlook 2016. Available: http://www.pbl.nl/sites/default/files/cms/publicaties/pbl-2016-national-energy-outlook-2016.PDF.
7. Government of the Netherlands. 2015. *Dutch agricultural exports top 80 billion Euros* [Online]. Available: https://www.government.nl/latest/news/2015/01/16/dutch-agricultural-exports-top-80-billion-euros.
8. Ministry for Climate Protection, E., Agriculture, Nature Conservation and Consumer Protection. 2011. The food industry in North Rhine-Westphalia. Quality and enjoyment from the regions. Available: https://www.umwelt.nrw.de/fileadmin/redaktion/Broschueren/broschuere_ernaehrung_english.pdf.
9. STOWA Delta Proof. 2014. Effect of climate change on agriculture. Available: http://deltaproof.stowa.nl/pdf/Effects_of_climate_change_on_agriculture_?rId=62.
10. European Commission. 2015. Nutrient pollution in Dutch streams is falling, but further reductions needed. *Science for Environmental Policy* [Online]. Available: http://ec.europa.eu/environment/integration/research/newsalert/pdf/nutrient_pollution_in_Dutch_streams_is_falling_402na2_en.pdf.
11. Netherlands Enterprise Agency. 2017b. MIA (Environmental investment rebate) and Vamil (Arbitrary depreciation of environmental investments).
12. City of Frankfurt. 2017c. *Green building FrankfurtRhineMain: Architecture award for sustainable construction* [Online]. Available: https://www.frankfurt.de/sixcms/detail.php?id=3077&_ffmpar%5B_id_inhalt%5D=17338690.

13. Environment Frankfurt. 2017. *What we are doing. Climate protection and energy supply* [Online]. Available: https://www.frankfurt-greencity.de/en/environment-frankfurt/climate-protection-and-energy-supply/what-we-are-doing/.
14. City of Frankfurt. 2013. Guidelines for economic building. Available: http://www.energiemanagement.stadt-frankfurt.de/Englisch/Guidelines-for-economic-building.pdf.
15. Investinwallonia. 2015. Wallonia Business & Competitiveness Clusters, Partners for your projects! Available: http://www.investinwallonia.be/publications/wallonia-business-competitiveness-clusters-partners-for-your-projects/?lang=en.
16. Ibid.
17. Ibid.
18. Ibid.
19. Netherlands Enterprise Agency. 2017a. *Energy investment allowance* [Online]. Available: http://english.rvo.nl/subsidies-programmes/energy-investment-allowance-eia.
20. Wageningen University and Research. 2017. *Water nexus* [Online]. Available: http://www.wur.nl/en/Expertise-Services/Chair-groups/Agrotechnology-and-Food-Sciences/Sub-department-of-Environmental-Technology/Water-Nexus.htm.
21. City of Frankfurt. 2017b. *Frankfurt saves electricity—Support programme for households* [Online]. Available: http://www.frankfurt.de/sixcms/detail.php?id=3077&_ffmpar%5B_id_inhalt%5D=17291768.
22. City of Frankfurt. 2017a. *Frankfurt saves electricity—Support programme for companies, associations and community centers* [Online]. Available: http://www.frankfurt.de/sixcms/detail.php?id=3077&_ffmpar[_id_inhalt]=17338692.
23. Sociaal-Economische Raad (Social and Economic Council). 2013. Energy agreement for sustainable growth: Implementation of the energy agreement. Available: https://www.ser.nl/en/publications/publications/2013/energy-agreement-sustainable-growth.aspx.
24. Environment Frankfurt. 2017. *What we are doing. Climate protection and energy supply* [Online]. Available: https://www.frankfurt-greencity.de/en/environment-frankfurt/climate-protection-and-energy-supply/what-we-are-doing/.
25. Biobased Delta. 2017. *Agro meets chemistry* [Online]. Available: http://biobaseddelta.nl/.

26. Duurzame Zuivelketen. 2017. Sustainable dairy chain: Vision, goals and approach. Available: http://www.duurzamezuivelketen.nl/files/vision-objectives-and-approach.pdf.

27. Duurzame Zuivelketen. 2012. *Energy scan for dairy farms* [Online]. Available: http://www.duurzamezuivelketen.nl/en/news/energy-scan-for-dairy-farms2.

28. Dutch Dairy Association. 2014. *Dutch dairy sector switches to sustainable soy* [Online]. Available: http://www.nzo.nl/en/dutch-dairy-sector-switches-to-sustainable-soy/.

29. Netherlands Enterprise Agency. 2014. Sustainable biomass production and use. Lessons learned from the Netherlands Programme Sustainable biomass (NPSB) 2009–2013. Available: http://www.biofuelstp.eu/downloads/netherlands/netherlands-sustainable-biomass.pdf.

30. Investinwallonia. 2015. Wallonia Business & Competitiveness Clusters, Partners for your projects! Available: http://www.investinwallonia.be/publications/wallonia-business-competitiveness-clusters-partners-for-your-projects/?lang=en.

31. Nutrient Platform NL. 2017. *About nutrient platform* [Online]. Available: https://www.nutrientplatform.org/en/about-nutrient-platform/.

References

Biobased Delta. 2017. *Agro meets chemistry* [Online]. http://biobaseddelta.nl/

City of Frankfurt. 2013. Guidelines for economic building. http://www.energiemanagement.stadt-frankfurt.de/Englisch/Guidelines-for-economic-building.pdf

———. 2017a. *Frankfurt saves electricity—Support programme for companies, associations and community centers* [Online]. http://www.frankfurt.de/sixcms/detail.php?id=3077&_ffmpar[_id_inhalt]=17338692

———. 2017b. *Frankfurt saves electricity—Support programme for households* [Online]. http://www.frankfurt.de/sixcms/detail.php?id=3077&_ffmpar%5B_id_inhalt%5D=17291768

———. 2017c. *Green building FrankfurtRhineMain: Architecture award for sustainable construction* [Online]. https://www.frankfurt.de/sixcms/detail.php?id=3077&_ffmpar%5B_id_inhalt%5D=17338690

Dutch Dairy Association. 2014. *Dutch dairy sector switches to sustainable soy* [Online]. http://www.nzo.nl/en/dutch-dairy-sector-switches-to-sustainable-soy/

Duurzame Zuivelketen. 2012. *Energy scan for dairy farms* [Online]. http://www. duurzamezuivelketen.nl/en/news/energy-scan-for-dairy-farms2

————. 2017. Sustainable dairy chain: Vision, goals and approach. http:// www.duurzamezuivelketen.nl/files/vision-objectives-and-approach.pdf

Environment Frankfurt. 2017. *What we are doing. Climate protection and energy supply* [Online]. https://www.frankfurt-greencity.de/en/environment-frankfurt/climate-protection-and-energy-supply/what-we-are-doing/

European Commission. 2015. Nutrient pollution in Dutch streams is falling, but further reductions needed. *Science for Environmental Policy* [Online]. http://ec.europa.eu/environment/integration/research/newsalert/pdf/nutrient_pollution_in_Dutch_streams_is_falling_402na2_en.pdf

Government of the Netherlands. 2015. *Dutch agricultural exports top 80 billion euros* [Online]. https://www.government.nl/latest/news/2015/01/16/dutch-agricultural-exports-top-80-billion-euros

ICPR. 2015. Strategy for the IRBD Rhine for adapting to climate change. http:// www.iksr.org/fileadmin/user_upload/Dokumente_en/Reports/219_en.pdf

————. 2017a. *Convention on the protection of the Rhine* [Online]. http://www. iksr.org/en/international-cooperation/legal-basis/convention/index.html

————. 2017b. *Industry* [Online]. http://www.iksr.org/en/uses/industry/index. html

Investinwallonia. 2015. Wallonia Business & Competitiveness Clusters, Partners for your projects! http://www.investinwallonia.be/publications/wallonia-business-competitiveness-clusters-partners-for-your-projects/?lang=en

Ministry for Climate Protection, Environment, Agriculture, Nature Conservation and Consumer Protection. 2011. The food industry in North Rhine-Westphalia. Quality and enjoyment from the regions. https://www. umwelt.nrw.de/fileadmin/redaktion/Broschueren/broschuere_ernaehrung_english.pdf

Netherlands Enterprise Agency. 2014. Sustainable biomass production and use. Lessons learned from the Netherlands Programme Sustainable Biomass (NPSB) 2009–2013. http://www.biofuelstp.eu/downloads/netherlands/netherlands-sustainable-biomass.pdf

————. 2017a. *Energy investment allowance* [Online]. http://english.rvo.nl/subsidies-programmes/energy-investment-allowance-eia

————. 2017b. MIA (Environmental investment rebate) and Vamil (Arbitrary depreciation of environmental investments).

NRW Invest. 2011. Energy in North Rhine-Westphalia. http://www.energieregion.nrw.de/_database/_data/datainfopool/Energy_in_North_Rhine-Westphalia_Facts_Figures.pdf

Nutrient Platform NL. 2017. *About nutrient platform* [Online]. https://www. nutrientplatform.org/en/about-nutrient-platform/

PBL Netherlands Environmental Assessment Agency. 2016. National energy outlook2016.http://www.pbl.nl/sites/default/files/cms/publicaties/pbl-2016-national-energy-outlook-2016.PDF

Sociaal-Economische Raad (Social and Economic Council). 2013. Energy agreement for sustainable growth: Implementation of the energy agreement. https://www.ser.nl/en/publications/publications/2013/energy-agreement-sustainable-growth.aspx

STOWA Delta Proof. 2014. Effect of climate change on agriculture. http://del-taproof.stowa.nl/pdf/Effects_of_climate_change_on_agriculture_?rId=62

Wageningen University and Research. 2017. *Water nexus* [Online]. http://www. wur.nl/en/Expertise-Services/Chair-groups/Agrotechnology-and-Food-Sciences/Sub-department-of-Environmental-Technology/Water-Nexus.htm

13

Best Practices

Introduction

The following best practices have been identified from the case studies of New York City and Singapore; Massachusetts and Ontario; Denmark and Korea; and the Colorado River Basin, Murray-Darling River Basin, and Rhine River Basin implementing a variety of fiscal and non-fiscal tools that create interdependencies and synergies between the nexus systems while reducing trade-offs between the systems in the development of a green economy. These best practices can be implemented by other locations around the world attempting to reduce water-energy-food nexus pressures and achieve a green economy.

13.1 Fiscal Tools

From the case studies, a variety of best practices have been identified in the use of fiscal tools to create interdependencies and synergies between the nexus systems while reducing trade-offs between the systems in the development of a green economy.

© The Author(s) 2018
R.C. Brears, *The Green Economy and the Water-Energy-Food Nexus*,
DOI 10.1057/978-1-137-58365-9_13

Market-Based Instruments and Pricing

A central aspect of encouraging green growth is integrating the natural asset base into everyday market decisions, which in turn reduces nexus pressures. This can be achieved with market-based instruments and pricing encouraging reductions in resource use and promoting conservation, directing profits towards resource conservation, as well as using instruments to reduce overall nexus pressures.

Pricing Resource Usage and Promoting Conservation

Market-based instruments and pricing can be used to reduce nexus pressures by reducing usage, increasing efficiency, and promoting conservation. For example, a city has revised upwards its pricing of water to meet both rising demand for water and increasing operational costs, while one country has priced water and wastewater to encourage industry to use water efficiently and reduce pollution from wastewater. Meanwhile, another jurisdiction has implemented a one-size-fits-all statewide water pricing structure in which all domestic customers pay the same water rate with the same structure applying to commercial customers too. In the context of reducing fossil fuel consumption and encouraging public transportation, one city has a pay-as-you-use tool to reduce congestion.

Directing Profits to Resource Conservation

By pricing resources, the profits can be directed towards resource conservation projects that reduce nexus pressures. For example, one location charges water users downstream to support upstream water protection projects, while a city has a buy-local-food-produce initiative that involves the profits being used to improve the economic viability of the local community and preserve water quality in the city's watershed region. Similarly, one jurisdiction enables schools to sell locally grown food with a portion of the profits returned to the farmers.

Permits and Tradeable Rights and Certificates to Reduce Nexus Pressures

Many jurisdictions are using permitting and resource rights trading as tools to reduce nexus pressures. For example, one location requires large water users to acquire a permit with the cost of the permit dependent on the risk to the environment. Meanwhile, another jurisdiction has developed a water market that is based on a cap-and-trade system with water distributed to users via water rights. In the context of reducing energy-related nexus pressures, one location mandates that all electricity companies must produce a minimum amount of renewable energy and if shortfalls occur they must purchase tradeable renewable energy certificates.

Environmental Taxes

Environmental taxes aim to raise the cost of production or consumption of environmentally damaging goods so as to limit their demand. Tax exemptions as well as incentives are commonly used to encourage the uptake of resource-efficient technology as well as encourage behavioural change. For instance, one city offers a property tax abatement for the installation of solar PV systems while another location has a biogas-from-agricultural waste tax exemption to support turning half of all livestock waste into biomass. Similarly, one country has created energy investment allowances that allow companies to make investments in energy efficiency with the allowance in the form of corporate tax deductions in addition to a scheme that enables companies to invest in environmentally friendly products or innovations with similar taxation benefits. Taxes can also be used to encourage pro-nexus decision-making; for example, one jurisdiction has a tax abatement for new buildings that meet a variety of voluntary international sustainability standards; one location offers cities the ability to provide property tax rebates for landowners who commit their land to urban agriculture; while another jurisdiction allows farmers to receive a tax credit for donating produce to community food programmes. Meanwhile, one jurisdiction charges farmers a tax for fertiliser use if they have not signed up to a fertiliser register that supports a pesticide reduction plan.

Financial Incentives

Financial instruments are often used to promote resource-efficient products and technologies.

Enhancing Resource Efficiency

In the context of reducing water-related nexus pressures, financial incentives can be used to increase the uptake of resource-efficient devices, appliances, and technologies. One location provides retrofit vouchers for residents to install water-efficient devices while another jurisdiction provides rebates for large water users to act on water-saving recommendations following a water audit, while one city provides funding for large water-using organisations to increase their water efficiency. Meanwhile, to reduce energy-related nexus pressures, jurisdictions are commonly providing funding for businesses to conduct energy audits as well as providing rebates for the installation of energy-efficient measures in buildings and facilities. Similarly, one jurisdiction not only provides grants to SMEs to increase their energy efficiency but also provides grants for training organisations to develop energy efficiency knowledge and skills in SMEs. Additionally, one city offers an expedited green permitting process with a fee reduction for new and renovated green buildings. Meanwhile, for domestic customers, jurisdictions use a variety of initiatives to reduce energy usage including cash bonuses to electricity consumers who conserve energy as well as rebates for the installation of energy-efficient appliances. Regarding reducing food-related nexus pressures and enhancing food productivity, grants are commonly distributed to farmers to conduct energy efficiency retrofits in addition to funding for agricultural-related renewable energy systems. Meanwhile another jurisdiction is offering a grant for urban irrigation water meters to be installed to track water usage while a state offers grants for the installation of water-efficient irrigation on agricultural land.

Resource Protection

Financial incentives can be used in a variety of ways to protect resources from over-consumption and reduce environmental impacts associated with the nexus. For instance, one state provides grants for the protection of water quality using innovative best-practice solutions. One case provides funding to crop producers to protect water and soil health as well as for projects that turn food waste into bioenergy and biofuel. To encourage consumers to purchase resource-friendly food produce, jurisdictions provide grants to increase knowledge of local food while another jurisdiction provides coupons to help low-income residents buy fresh, locally grown produce. In a similar fashion, another location provides grants for reducing the costs of selling food produce that uses alternatives to pesticides and other chemicals. Finally, one state provides grants for food and beverage customers to reduce and/or improve the quality of their waste for reuse.

Enhancing Infrastructure and Technology

Financial incentives can be used to encourage the development of infrastructure and technologies that reduce nexus pressures. For example, jurisdictions often provide loans to start-ups and SMEs that are developing water-related technologies including innovative water technologies that increase efficiency in wastewater treatment plants. To reduce energy consumption and increase renewable energy, one location provides funding for community and commercial wind projects, grants for the development of organic waste-to-energy projects, and funding for farms to develop renewable energy systems, while another location provides a home subsidy programme for the installation of renewable energy systems in houses. Governments are also providing funding to public and private sector actors to develop large-scale EV charging station networks. Regarding reducing food-related nexus pressures, one jurisdiction provides funds for agricultural producers to increase on-farm productivity by adopting hi-tech systems.

Payments for Ecosystem Services

To encourage private landowners to take actions that enhance ecosystem services for the benefit of all, one state provides payments to farmers who permanently restrict the use of agricultural land for non-farming purposes with payments also made for actions that reduce impacts on water quality. Another location provides a supplement to farmers who are converting their land to organic farming. Meanwhile, one state provides funding for projects that protect at-risk agricultural lands from conversion to urban development. Finally, one jurisdiction provides incentives for private landholders to manage their land in ways that conserve and enhance the environment.

13.2   Non-fiscal Tools

From the case studies a variety of best practices have been identified in the use of non-fiscal tools to create interdependencies and synergies between the nexus systems while reducing trade-offs between the systems in the development of a green economy.

Regulations

Regulations are often used to influence green growth by encouraging production efficiency and reducing the amount of wastage. The most commonly used regulatory tools include mandating resource conservation and benchmarking resource usage.

Resource Conservation

An array of jurisdictions have implemented resource conservation and efficiency requirements related to water, energy, and food. Regarding water efficiency, one city requires that during a drought all large water users need to conduct a self-audit and develop a conservation plan. Similarly, one location requires all large water users to submit water efficiency plans that include actions to save water while another jurisdiction has an Activities Decree that contains rules for companies to follow when using water. Regarding energy efficiency and conservation, a variety of tools are used, for instance, a regulation requiring all existing buildings undergoing green building award assessments having to achieve minimum energy efficiency standards; the requirement that all energy-intensive companies appoint an energy manager and submit energy efficiency plans; and that all large enterprises carry out energy audits every few years. Meanwhile, to enhance energy conservation in the built environment, a range of resource conservation regulations have been used by a range of jurisdictions including the requirement that: all renovated buildings meet the current state energy code, all commercial buildings to upgrade to energy-efficient lighting and install sub-metering within a set time frame, and all new homes comply with the highest building standards of the national construction code. Resource conservation regulations have also been placed on food-related issues; for instance, one city has initiated a law that provides for voluntary kerbside organic waste collection as well as mandating that all large-scale commercial food establishments separate organic waste.

Benchmarking Resource Usage

One city requires private buildings of a specific size or larger to measure their energy and water usage and report it to the city, enabling comparisons to be made between buildings. Similarly, another jurisdiction requires that all private buildings report their energy and water usage annually with a subset of the data available for the public, including the financial and property market, to compare and value buildings.

Standards and Mandatory Labelling

Jurisdictions use a variety of standards and mandatory labelling schemes to eliminate unsustainable products and practices from the market, with the most common being advanced built environment standards and resource efficiency labelling.

Advanced Building Codes

One state has building codes that meet or exceed international standards with a stretch code available for cities and towns to choose more energy-efficient codes. Meanwhile another jurisdiction mandates that all new or renovated buildings include a minimum amount of renewable energy in addition to ensuring that eco-friendly homes make available their energy consumption on an Internet portal.

Resource Efficiency Labelling

One jurisdiction requires certain products to be labelled with their water efficiency while another has mandated that all manufacturers and importers produce and sell energy-efficient products, with their efficiency levels clearly labelled. Another location requires that all buildings that use energy to regulate the indoor climate have energy labels before they are sold or let. One state mandates that all electricity suppliers must disclose to consumers the energy resources used.

Public Education and Skills Development

Public education can affect green economy innovation by increasing the acceptance of technological innovation as well as increasing capacity for future development of green economy technologies. This can be achieved through a range of initiatives including community education, on-the-job-training, and professional certification.

Community Education

One jurisdiction financially supports climate change education among Indigenous communities while another provides support to schools to implement sustainable energy programmes with schools receiving technical assistance and coaching. Similarly, another state helps schools save energy, conserve water, and develop sustainability-related skills.

On-the-Job Training

One city provides agricultural and food-related businesses employee training opportunities to increase productivity. Another jurisdiction has a learn-and-earn programme, where high school students are employed and receive clean energy training; meanwhile, another location offers clean energy internships with interns working in renewable energy and energy efficiency companies.

Professional Certification

One city is training green building specialists to increase the number of buildings achieving green certification. Another jurisdiction supports green workforce development programmes that offer courses and certificates. Meanwhile, one state trains low- and moderate-income-level women in energy-related sales and business management roles.

Public–Private Partnerships

Public–private partnerships (PPPs) are being increasingly recognised as offering feasible solutions to complement or replace public responsibilities for infrastructure and green economy–related services, with a variety of PPPs seen throughout the case studies including PPPs on developing infrastructure, public–private waste challenges, partnerships to protect resources, and PPP guidance for reducing nexus pressures.

PPPs on Developing Resource-Efficient Infrastructure

A variety of PPP models exist for developing resource-efficient infrastructure. For example, one jurisdiction has initiated PPPs in the development of EV charging station networks, while another location has developed a PPP in the development of bio-based fuels as well as biological processes for treating wastewater. Another location has made it mandatory for municipalities to cooperate with the private sector when implementing urban renewal projects.

Public and Private Sector Waste Challenges

One unique form of PPP is jurisdictions challenging the private sector to reduce their resource usage. For example, one city provides support to hotels and restaurants who voluntarily reduce water usage. The same city also challenges restaurants to reduce food waste and businesses to reduce their GHG emissions. Meanwhile, another jurisdiction challenges all non-domestic water-using customers to reduce their water usage with the city's utility providing expertise on how to do so.

Partnerships to Protect Resources

One jurisdiction has mandated that local conservation authorities can help protect waterways while the same jurisdiction assists landowners in protecting drinking water supplies. Meanwhile another location has established eco-friendly partnerships in which SMEs provide eco-friendly products in exchange for environmental management knowledge. Another state works with farmers to increase productivity of their soils. One city is even working with a watershed council to help better manage on-farm water quality and sell locally produced food that protects water quality. Finally, one country offers its cities deals in which the public sector will support cities and the private sector in developing innovative urban solutions to societal issues as well as green deals in which the public and private sectors work together to implement sustainability plans.

PPP Guidance

To facilitate the development of PPPs and reduce nexus pressures, one jurisdiction has created a renewable energy advocate that works with communities and industry to facilitate the development and generation of renewable energy, while another has created guidelines to reduce environmental impacts of new construction and renovation projects under public–private contracts. Finally, one country has developed a PPP on developing innovative solutions to water problems in the agricultural sector as well as sharing best practices.

Stakeholder Participation

Stakeholder participation is a condition for making resource management decisions more effective. There are a variety of stakeholder models being implemented throughout the case studies including community participation, stakeholder management of resources, and partnerships to develop green economy–related tools.

Facilitating Participation

One city promotes the inclusion of a garden in every public school to encourage young people to become actively involved in the production of local food. The same city also works with its housing association to provide residents with opportunities to participate in community garden initiatives, while another city has developed an urban farm plan to promote the development of inner city farms that involve numerous stakeholders.

Stakeholder Management of Resources

One particular jurisdiction has created committees that identify existing and future risks to drinking water sources while another location has implemented a water knowledge programme that involves multiple stakeholders conducting research on water-related issues. With regard to reducing food-related nexus pressures, one state has a legislatively mandated volunteer council that makes recommendations on how to strengthen local food access.

Partnerships to Develop Green Economy–Related Tools

There are examples of public agencies working with non-governmental organisations and researchers on reducing resource use and creating a green economy. For example, one city is working with multiple stakeholders on developing a carbon footprint assessment tool as well as working with industry to review various building codes to increase indoor environmental quality. One jurisdiction is providing financial support to an organisation that helps SMEs reduce their environmental impacts.

Information and Awareness-Raising

Information and awareness-raising initiatives are commonly used to promote green growth–related consumption choices, for example, purchasing environmentally friendly goods and services or using resources wisely.

Information and Technical Guides

Across the cases, there are many forms of informational and technical guides developed to reduce nexus pressures and encourage green growth. For example, one city provides free technical and advisory services to building owners wishing to go 'green', while another jurisdiction has developed guidelines to show that conservation of historical buildings and climate protection are not mutually exclusive. To enhance industrial efficiency, one location provides guidelines on how industry can be water efficient. There are also examples of public agencies providing energy and water efficiency information on consumer goods and appliances as well as making available renewable energy benefits. At the household level, one state provides home energy action kits for houses to reduce their energy and water usage while another location has enabled financial institutions to provide energy renovation advice to homeowners. A city even provides a service where a city representative visits households to help them save water and install water-efficient devices. Meanwhile, one city provides free advice to non-profits on how to reduce their water and energy usage, which in turn increases their resources available to help the needy. Regarding reducing food-related nexus pressures, one jurisdiction has developed an Internet portal that provides advice to farmers and companies to adapt to climate change while another state provides agricultural producers with free energy audits and technical support services to reduce their energy use. Finally, another state provides a best management practice guide for farmers interested in reducing energy usage and installing renewable energy systems.

Monitoring Resource Consumption

Jurisdictions have implemented a variety of initiatives to monitor resource consumption, for instance, enabling customers to view their water usage, enabling leaks to be detected easily, and developing a water benchmark that enables businesses to compare their water usage with other similar businesses. Similarly, one jurisdiction enables customers to access their own electricity data to reduce consumption. Meanwhile, one state has an agricultural certification programme that helps farmers evaluate their operations to make environment and productivity decisions.

School Education

School education is one of the most powerful tools for providing individuals with the skills and competencies to become sustainable consumers. A variety of initiatives have been implemented in the case studies, with the majority involving hands-on learning experiences.

Green Hands-On Experience

One jurisdiction provides young children with first-hand experience in sustainability-related issues in addition to including water conservation in the school syllabus, while another location has developed a service that provides education to children on nature and sustainable development. Similarly, another location embeds general sustainability into school curricula. One state has even developed a climate kids project that introduces young people to the basics of renewable energy through hands-on experiments.

Voluntary Labelling

One of the most common tools for influencing sustainable consumer choices is voluntary labelling of products and services, including green building certification and labelling of products' resource efficiency as well as their environmental protection qualities.

Green Building Certification

One city has a green building certification programme that has been expanded beyond buildings to include infrastructure and retail space. Meanwhile, a state has piloted a building and home rating and labelling programme that enables comparisons to be made between buildings and homes.

Resource Efficiency Certification

One city has a water efficiency building certification programme for non-domestic buildings while another location provides carbon labelling for tap water produced with minimal amounts of GHG emissions. One jurisdiction has created a high-efficiency energy label for products that voluntarily exceed government efficiency standards.

Resource Protection Labelling

One state has created a food label that identifies local products that use sustainable methods, while another location is testing the development of a sustainable land development certification scheme for construction sites; in addition, the same location has an organic logo that can be applied only to farms authorised for organic production. Meanwhile another location has created a food product label for consumers to identify high-quality, local food.

Clustering Policies

Numerous jurisdictions have created industrial clusters to reduce nexus pressures and enhance green growth. For instance, one state has developed a water technology cluster that aims to accelerate the development of water technologies. Similarly, a couple of locations have developed water networks that aim to share knowledge and collaborate on water-related issues as well as accelerate the testing of new technologies. Correspondingly, a country has developed a water nexus platform for the private sector and institutions to work together on developing solutions to water scarcity. In the context of energy, one state has developed energy clusters in areas including biomass, solar, and wind, while a city has created a green building cluster that increases the deployment of energy efficiency technologies in buildings. Finally, to reduce food-related nexus pressures, one government has formed a cluster that aims to increase food quality and food sector productivity.

Resource Mapping

Resource mapping is important in identifying resource efficiency and nexus-related opportunities for investment as well as facilitating stakeholder engagement and raising public awareness on ongoing projects and their benefits. Resource mapping comes in numerous forms including interactive tools as well as performance and industry mapping.

Interactive Tools

One city provides a free interactive tool to help residents estimate their buildings' solar potential, while a state provides government buildings with real-time energy tracking. Meanwhile, another jurisdiction provides interactive water source protection area maps. Similarly, another location has developed an online tool for municipalities and property owners to map climate change vulnerabilities. One state has developed a geothermal map that provides municipal-level information on the percentage of actual heat required that could be met by geothermal resources.

Performance and Industry Mapping

One state has mapped its water industry, identifying strategies to increase the industry's impact in reducing water-related nexus pressures, while one city provides a map to show how their buildings compare with others in terms of energy and water usage. Meanwhile, another location maps food services and retailers that use local produce in their ingredients.

Public Procurement

Governments are frequently the largest consumer of goods and services and so they have the power to influence markets towards sustainability through the quantity of their purchases while providing good sustainable consumption examples to their citizens. For instance, one city has made the procurement of local and fresh food a priority in its school system, while another jurisdiction is increasing local food procurement throughout the public sector, while another location uses green procurement policies to promote energy-efficient appliances. Finally, public agencies have facilitated the adoption of water technologies in public facilities while another location enables small businesses that focus on innovative solutions to submit tenders for government procurement contracts.

Demonstration Projects

Demonstration projects are frequently undertaken to illustrate the feasibility and commercial viability of green economy initiatives and showcase the numerous socio-economic benefits to a variety of stakeholders. Demonstration projects are often undertaken as part of a test-bedding initiative or via public demonstration projects.

Test-Bedding

One city engages vendors to test their energy solutions in city-owned buildings while another city funds the testing of smart and sustainable solutions to generate local performance data for verification for eventual commercialisation. Another location has established a project that enables companies to test smart energy grid infrastructure and technologies as well as provide locations for small businesses to conduct water technology testing, with well-performing technologies receiving special contracts for their products.

Public Demonstrations

One city has initiated a demonstration to show how organic waste from food outlets can be used to generate energy while another location is demonstrating a smart water grid that enables real-time monitoring of water quantity and quality. One state has a public demonstration project that shows how homes of the future will be low-carbon in addition to providing village communities support on climate protection initiatives that will make them energy-savvy.

Awards and Public Recognition

Awards and public recognition schemes facilitate the development of role models who strengthen social norms and guide society on making decisions that reduce nexus pressures and enhance green growth. These schemes come in a variety of forms including school awards, recognition of outstanding sustainability initiatives as well as public competitions.

School Awards

One city has an annual art and poetry contest for schoolchildren on water-related issues, while another jurisdiction holds water contests for schools with students submitting essays, videos, or posters. Meanwhile, one state has a climate protection competition involving day care centres with the best initiatives celebrated.

Outstanding Sustainability Initiatives

One city recognises outstanding achievements in embedding sustainability in projects and buildings while another recognises builders and planners who make exceptional contributions to architecture and climate protection. Meanwhile, another location awards businesses for their commitment to sustainability and real improvements made in the process. Finally, one location has developed an awards system that recognises all members of society for excellence in environmental sustainability.

Public Competitions

One state holds a cleanweb haccelerator in which contestants build the best clean energy mobile app and online technology.

Voluntary Agreements

Voluntary agreements facilitate group actions that aim to improve the productivity and environmental sustainability of industry or a particular economic sector. One jurisdiction has a voluntary partnership for companies to become more energy efficient, with participants receiving advice and support. Similarly, another location has established voluntary agreements between a company, or group of companies from the same industrial sub-sector, and the government to implement energy management systems and undertake energy audits in exchange for energy tax relief. Meanwhile, in one location, organisations from the public and private sectors as well as unions signed an agreement to increase energy efficiency and renewable energy. To reduce food-related nexus pressures, one country has developed an agenda with its agri-food industry to undergo a complete sustainable transition while another country has approved an industry alliance to make binding agreements for all its members and work towards climate neutrality and environmental protection.

Conclusions

The traditional economic model of employing various types of capital, including human, technological, and natural, to create goods and services has led to higher living standards and improved human well-being. However, this growth has come at the cost of environmental degradation. At the same time, the global economy is under various stresses from rising population and economic growth, rapid urbanisation, soil and land degradation, air pollution, rising inequality, and so forth in addition to climate change. In response, many multi-lateral organisations have called for the development of a green economy that improves human well-being, enhances social equity, and reduces environmental degradation. While the definition of the term 'green economy' differs between organisations, the term overall seeks to reduce environmental degradation and spur green economic growth, all the while ensuring social equity.

A key challenge to achieving a green economy is managing the water-energy-food nexus; however, increasing water scarcity, rising demand for energy, and the need to reduce fossil fuel usage as well as feeding an increasing population is leading to potential water, energy, and food insecurity. This in turn will cause social and geopolitical tension in addition

© The Author(s) 2018
R.C. Brears, *The Green Economy and the Water-Energy-Food Nexus*,
DOI 10.1057/978-1-137-58365-9

to irreparable environmental degradation. As such, identifying the interactions across, and in between, the nexus sectors is key to developing a green economy that improves human well-being and environmental sustainability for both current and future generations. In fact, a green economy that reduces nexus pressures will enjoy a multitude of benefits including increased resource productivity, reduced waste, increased technological development, productive ecosystem services, enhanced poverty alleviation through green growth, and greater awareness on the environmental and nexus interactions.

The transition towards the green economy will require policy instruments that reduce water-energy-food nexus pressures. In particular, policies will be required to promote social and technological innovations that increase resource efficiency and conservation in order to reduce nexus pressures. Policies that reduce nexus pressures in the green economy can be implemented by fiscal and non-fiscal tools. Fiscal tools include market-based instruments and pricing that integrate the natural capital base into every market decisions through levies, charges, tradeable permits, and so forth. This enables governments to achieve natural resource consumption targets, provide incentives for the development of new technologies, allocate natural resources to parties that value them the most, and so on. Environmental taxes are used to raise the cost of production or consumption of environmentally damaging goods with environmental taxes commonly levied on energy and water. Fiscal incentives including subsidies and grants facilitate the adoption of green technologies and practices and encourage sustainable consumption choices. PES instruments are used to create a marketplace for ecosystem services in which private landowners are paid to implement environmentally friendly actions which benefit specific or general ecosystem service users.

A variety of non-fiscal tools can be used to promote the development of green growth–related technologies and services that reduce nexus pressures as well as enhance the public's awareness and understanding of natural resources. While regulations can be used to control behaviour, they can also be used to encourage pro-environmental behaviour such as encouraging the purchasing of resource-efficient products and devices. Standards and mandatory labelling schemes aim to limit environmental damage from products when they are consumed or used, with the ultimate aim of eliminating unsustainable products from the market. Public education and skills development programmes are key to enhancing the green economy as increased knowledge facilitates the acceptance of green technology innovations by society as a whole; well-educated, skilled workers are required to develop innovative green technologies; and people with higher education and skills find it easier to adopt some green economy innovations. PPPs are an important tool in enhancing the green economy as they enhance investment in innovative, resource-related services such as energy and water systems. Stakeholder participation is a key tool in managing nexus pressures as it facilitates exchanges and debates on new ideas that can reduce resource scarcity pressures. Information and awareness-raising initiatives are commonly used to promote sustainable consumption of resources including the purchasing of environmentally friendly goods and services. School education programmes are a powerful tool for providing individuals with the skills to become sustainable consumers. Voluntary labelling is a common tool for influencing sustainable consumption choices with most eco-labelling schemes verified by third-parties including governments and non-governmental organisations. Clustering policies are frequently developed to concentrate skill sets and resources on particular issues such as developing renewable energy systems. Resource maps are commonly developed to identify opportunities for targeting investments as well as to raise public awareness on ongoing environmental-related projects and their benefits. Public procurement policies can influence the market towards sustainability due to the sheer volume of government-related purchases, which in turn can provide an example to consumers of purchasing environmentally responsible products and services. Demonstration projects are often used to illustrate the feasibility and commercial viability of green economy initiatives and showcase their socio-economic benefits to a wide array of stakeholders. Awards and public recognition schemes strengthen social norms with regard to the environment and natural resources by showcasing sustainable actions, resulting in others likely to shift their behaviours to feel more connected to their role models. Voluntary agreements can be formed between the government and private sector on reducing resource consumption and increasing resource efficiency. Because the agreements are public, signatories are likely to improve resource efficiency and environmental performance beyond the level required by existing environmental legislation and regulations.

From the case studies of New York City, Singapore, Massachusetts, Ontario, Denmark, Korea, the Colorado River Basin, the Murray-Darling River Basin, and the Rhine River Basin implementing a variety of fiscal and non-fiscal tools that create interdependencies and synergies between the nexus systems while reducing trade-offs between the systems in the development of a green economy, the following best practices have been identified for other locations around the world attempting to reduce water-energy-food nexus pressures and achieve a green economy.

Market-based instruments and pricing can be used to reduce nexus pressures by reducing usage, increasing efficiency, and promoting conservation. For example, jurisdictions have increased the price of water to ensure it covers rising economic costs as well as reduce demand while other locations have charged for resource usage, for example, a pay-as-you-use tool for reducing road congestion and encouraging public transportation. By pricing resources, profits can be directed towards resource conservation projects that reduce nexus pressures. For example, water charges for downstream users are used to support upstream water protection projects while profits from local food produce are used to preserve water quality in a city's watershed region.

Jurisdictions are using permitting and resource rights trading as tools to reduce nexus pressures, with one location requiring large water users to acquire permits that increase in cost as environmental risk increases, while another jurisdiction has developed a water market that is based on a cap-and-trade system with water distributed to users via water rights. Electricity companies can also purchase tradeable renewable energy certificates to meet minimum renewable energy production requirements.

Tax exemptions as well as incentives are commonly used to encourage the uptake of resource-efficient technology, for example, a property tax abatement for the installation of solar PV systems and a waste-to-energy tax exemption for agricultural sites. Corporate tax deductions are also provided for investments in resource-efficient technologies and environmentally friendly products or innovations. Taxes can also be used to encourage sustainable behaviours including meeting voluntary international building sustainability standards, committing land to urban agriculture, or donating produce to community food programmes.

Financial incentives are used to increase the uptake of resource-efficient devices, appliances, and technologies; for example, one location offers vouchers for installing water-efficient devices while another location offers rebates for large water users to install water-efficient products and technologies. Meanwhile, jurisdictions are providing funding for businesses to conduct energy audits as well as providing rebates for the installation of energy-efficient measures. Grants are commonly distributed to farmers to conduct energy-efficient retrofits in addition to funding for agricultural-related renewable energy systems and water-efficient irrigation systems. Financial incentives can be used in a variety of ways to protect resources from over-consumption and reduce environmental impacts, with grants for water quality protection frequently offered. Financial incentives can also be used to encourage the development of infrastructure and technologies that reduce nexus pressures. For example, governments are providing funding to public and private sector actors to develop large-scale resource efficiency projects such as EV charging station networks.

Regarding PES markets, payments are made to farmers who permanently restrict the use of agricultural land for non-farming purposes that also protect water quality as well as payments for converting agricultural land to organic farming.

An array of jurisdictions have implemented resource conservation and efficiency requirements related to water, energy, and food. Regarding water, examples include large water users needing to: conduct self-audits during droughts, submit water efficiency plans that include actions to save water, and abide by rules when using water. Regarding energy efficiency and conservation, a variety of tools are used, for instance requiring all existing buildings undergoing green building award assessments having to achieve minimum energy efficiency standards, and mandating that all energy-intensive companies appoint an energy manager and submit energy efficiency plans. Regulations have also been placed on food-related issues, for instance, mandating that all large-scale commercial food establishments separate organic waste. The use of regulatory benchmarking of energy and water use in buildings has also been implemented with one city requiring private buildings of a specific size or larger to measure their energy and water usage and report it to the city, enabling comparisons to be made between buildings.

Jurisdictions use a variety of standards and mandatory labelling schemes to eliminate unsustainable products and practices from the market, with the most common being advanced built environment standards, such as requiring all new or renovated buildings include a minimum amount of renewable energy, as well as resource efficiency labelling for water- and energy-using appliances.

Public education can affect green economy innovation by increasing the acceptance of technological innovations as well as increasing capacity for the future development of green economy technologies; for example, one jurisdiction provides support to schools to implement sustainable energy programmes with schools receiving technical assistance and coaching. Meanwhile, other locations provide on-the-job training for employees on green economy–related issues as well as clean energy internships for high school students. Additionally, another location is offering training programmes to enable low- and moderate-income-level women to enter energy-related sales and business management roles.

A variety of PPP models exist for developing resource-efficient infrastructure. For example, one jurisdiction has initiated PPPs in the development of EV charging station networks, while another location has developed a PPP in the development of bio-based fuels as well as biological processes for treating wastewater. PPPs have also been developed to challenge the private sector to reduce resource usage with public agencies providing support and training for participating companies. Another form of PPPs is knowledge partnerships, for instance, one location is transferring knowledge on environmental management systems in exchange for private sector eco-friendly products.

There are a variety of stakeholder models being implemented throughout the case studies including community participation, stakeholder management of resources, and partnerships to develop green economy–related tools. For example, one city works with its housing association to provide residents with opportunities to participate in community garden initiatives; another location has created a legislatively mandated volunteer council to make recommendations on how to strengthen local food access, while another jurisdiction is working with industry to review building codes to increase indoor environmental quality.

Information and awareness-raising initiatives are commonly used to promote green growth–related consumption choices, for example, purchasing environmentally friendly goods and services or using resources wisely. For example, one city provides free technical and advisory services to building owners wishing to go 'green', while another jurisdiction has developed guidelines to show that conservation of historical buildings and climate protection are not mutually exclusive, while another location provides free advice to non-profits on how to reduce their water and energy usage. Finally, one location provides a best management practice guide for farmers interested in reducing energy usage and installing renewable energy systems.

Jurisdictions have also implemented a variety of initiatives to enable stakeholders to monitor their resource consumption, for instance enabling water customers to view their water usage, developing a water benchmark that enables businesses to compare their water usage with other similar businesses, and implementing an agricultural certification programme that helps farmers evaluate their operations to make environment and productivity decisions.

School education is one of the most powerful tools for providing individuals with the skills and competencies to become sustainable consumers. A variety of initiatives have been implemented that involve hands-on learning experiences; for example, one jurisdiction provides young children with first-hand experience in sustainability-related issues in addition to including water conservation in the school syllabus, while another location teaches young people the basics of renewable energy through hands-on experiments.

One of the most common tools for influencing sustainable consumer choices is voluntary labelling of products and services. For instance, one location has created a high-efficiency energy label for products that voluntarily exceed government efficiency standards. Regarding food, organic food logos have been created that can be applied only to farms authorised for organic production.

Numerous jurisdictions have created industrial clusters to reduce nexus pressures and enhance green growth. For instance, one state has developed a water technology cluster that aims to accelerate the development of water technologies, while another state has developed energy clusters in areas including biomass, solar, and wind. Meanwhile one government has formed a cluster that aims to increase food quality and food sector productivity.

Resource mapping is important in identifying resource efficiency and nexus-related opportunities. This enhances investment, facilitates stakeholder engagement as well as raises public awareness on ongoing projects and their benefits. Mapping tools come in a variety of forms including interactive tools that help residents estimate their buildings' solar potential and maps that highlight sustainable behaviour, for instance mapping food services and retailers that use local produce in their ingredients.

Governments are frequently the largest consumer of goods and services and so they have the power to influence market transitions towards sustainability through the quantity of their purchases while providing good sustainable consumption examples to their citizens, which include using green procurement policies to promote energy-efficient appliances.

Demonstration projects are frequently undertaken to illustrate the feasibility and commercial viability of green economy initiatives and showcase the numerous socio-economic benefits to a variety of stakeholders. Demonstration projects are often undertaken as part of test-bedding initiatives, such as enabling vendors to test their energy solutions in city-owned buildings. Meanwhile, public demonstration projects showcase the viability of green technologies, for example, how organic waste from food outlets can be used to generate energy.

Awards and public recognition schemes facilitate the development of role models who strengthen social norms and guide society towards making decisions that reduce nexus pressures and enhance green growth, examples of which include annual art and poetry contests for schoolchildren on water-related issues, recognising builders and planners who make exceptional contributions to architecture and climate protection, and awards for businesses that make substantial commitments to sustainability.

Voluntary agreements facilitate group actions that aim to improve the productivity and environmental sustainability of an industry, examples of which include voluntary partnerships for companies to become more energy efficient, with participants receiving government advice and support, while another location has approved a food industry alliance to make binding agreements for all its members and work towards climate neutrality and environmental protection.

In conclusion, reducing water-energy-food nexus pressures and developing a green economy is not a static activity. Instead it requires a variety of fiscal and non-fiscal tools to encourage green technologies and practices that create interdependencies and synergies between the nexus systems while reducing trade-offs between the systems.

Index[1]

[1]Note: Page number followed by 'n' refers to end notes.

© The Author(s) 2018
R.C. Brears, *The Green Economy and the Water-Energy-Food Nexus*,
DOI 10.1057/978-1-137-58365-9

food security, 24, 27, 39, 42,
    291
food-water. *See* water-energy-food
    nexus
funding, 55, 95, 123, 128, 129, 133,
    135–7, 149–51, 153, 154,
    160–5, 167–70, 183, 184,
    188, 190–3, 199, 201, 202,
    204, 205, 210, 217, 227, 281,
    285, 289, 291, 292, 294,
    297, 298, 322, 327, 330,
    332–3, 339, 362, 364,
    388–90, 407
funds, 115, 123, 124, 161, 168, 183,
    184, 193, 201, 204, 205,
    217–19, 238, 252, 253, 330,
    332, 339

Germany
    Blue Angel label, 64
    life cycle costing, 68
globalisation, 5
GPP. *See* green public procurement
    (GPP)
grants, 55, 96, 129, 149, 188, 217,
    252, 324, 360, 404
Green Credits, 252, 265
green economy
    ecosystem resilience, 8
    energy security, 26, 27
    fiscal tools, xii, 51–7
    food security, 27
    non-fiscal tools, xii, 57–70
    objectives, 8–11

resource-use efficiency,
    8, 10
social equity, 7, 11
water security, 24, 25
green growth. *See* green
    economy
green procurement, 68, 219, 220,
    259, 267, 400, 410
green public procurement (GPP),
    219

H
health bucks, 3
Hong Kong
    Mandatory Energy Efficiency
    Labelling Scheme, 59

I
incentives, xii, 40–2, 52, 54–5,
    91–3, 128, 153, 154, 161,
    168, 169, 189–91, 204, 229,
    252, 280–2, 284, 285, 289,
    290, 296, 297, 313, 357,
    387–90, 404, 407
inequality. *See* Poverty
information and awareness-raising,
    62–3, 396–7, 405, 409
innovation, xi, xii, xiii, 6, 8, 40,
    59, 137, 151, 169, 203, 204,
    218, 282, 291, 332, 353,
    362–4, 371, 387, 392,
    404–8
Ireland, 54
    One Good Idea, 69–70

Printed in the United
by Billing & Sons

Printed in the United States
By Bookmasters